Organic Synthesis

State of the Art 2003–2005

Organic Synthesis

State of the Art 2003–2005

Douglass F. Taber

University of Delaware
Newark, DE

WILEY-INTERSCIENCE

A JOHN WILEY & SONS, INC., PUBLICATION

Published by John Wiley & Sons, Inc., Hoboken, New Jersey.
Published simultaneously in Canada.

For general information on our other products and services or for technical support, please contact our Customer Care Department within the United States at (800) 762-2974, outside the United States at (317) 572-3993 or fax (317) 572-4002.

Wiley also publishes its books in a variety of electronic formats. Some content that appears in print may not be available in electronic format. For information about Wiley products, visit our web site at www.wiley.com.

Library of Congress Cataloging-in-Publication Data is available.

ISBN-13 978-0-470-05331-7
ISBN-10 0-470-05331-3

Printed in the United States of America.

10 9 8 7 6 5 4 3 2 1

Contents

Preface

Starting in January of 2003, I have been publishing a weekly Organic Highlights column (http://www.organic-chemistry.org/). Each column covers a topic, pulling together the most significant developments in that area of organic synthesis over the previous six months. All of the columns published in 2004 and 2005 are included in this book.

So, why this book, if the columns are already available free on the Web? First, there are a lot of them, 103 in this book. It is convenient having them all in one place. Too, there is an index of senior authors, and a subject/transformation index. Most important, this collection of columns, taken together, is an effective overview of the most important developments in organic synthesis over the two-year period. The dates have been left on the columns in this volume, so they will be easy to locate on the Web. The web columns include electronic links to the articles cited.

There are journals that publish abstracts and/or highlights. The columns here differ from those efforts in that these columns take the most important developments in an area, e.g. the Diels-Alder reaction, over a six-month period, and put them all together, with an accompanying analysis of the significance of each contribution.

The first column of each month is devoted to a total synthesis. So many outstanding total syntheses appear each year, no attempt was made to be comprehensive. Rather, each synthesis chosen was selected because it contributed in some important way to the developing concepts of synthesis strategy and design. It is important to note that even if a total synthesis was not featured as such, all new reaction chemistry in that synthesis was included at the appropriate place in these Highlights.

I recommend this book to the beginning student, as an overview of the state of the art of organic synthesis. I recommend this book to the accomplished practitioner, as a handy reference volume covering current developments in the field.

These Highlights are primarily drawn from the *Journal of the American Chemical Society, Journal of Organic Chemistry, Angewandte Chemie, Organic Letters, Tetrahedron Letters,* and *Chemical Communications.* If you come across a paper in some other journal that you think is worthy of inclusion, please send it to me!

DOUGLASS F. TABER

Department of Chemistry and Biochemistry
University of Delaware
Newark, DE 19716
taberdf@udel.edu
http://valhalla.chem.udel.edu
(302) 831-2433
Fax: (302) 831-6335

Transition Metal Mediated Reactions in Organic Synthesis

January 12, 2004

This week's Highlights focuses on three transition metal-catalyzed reactions. Jin-Quan Yu of Cambridge University reports (Organic Lett. 2003, 5, 4665-4668) that Pd nanoparticles catalyze the hydrogenolysis of benzylic epoxides. The reaction proceeds with inversion of absolute configuration ($1 \rightarrow 2$).

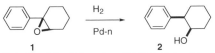

Laurel Schafer of the University of British Columbia reports (Organic Lett. 2003, 5, 4733-4736) that terminal alkynes undergo smooth hydroamination with a Ti catalyst. The intermediate imine **4** can be hydrolyzed to the aldehyde **5** or reduced directly to the amine **6**. The alkyne to aldehyde conversion has previously been carried out by hydroboration/oxidation (J. Org. Chem. 1996, 61, 3224), hydrosilylation/oxidation (Tetrahedron Lett. 1984, 25, 321), or Ru catalysis (J. Am. Chem. Soc. 2001, 123, 11917). There was no previous general procedure for the anti-Markownikov conversion of a terminal alkyne to the amine.

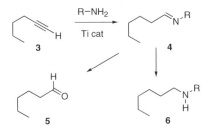

The construction of enantiomerically-pure carbocycles is a general problem in organic synthesis. Dirk Trauner (UC Berkeley) reports (Organic Lett. 2003, 5, 4113-4115) an elegant intramolecular Heck cyclization. The alcohol **7** is readily prepared in enantiomerically-pure form. Conditions can be varied so that either **8** or **9** is the dominant product from the cyclization.

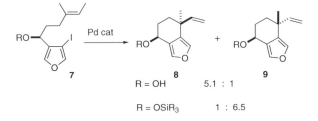

R = OH   5.1 : 1

R = OSiR$_3$ 1 : 6.5

Biocatalytic Asymmetric Hydrogen Transfer

January 19, 2004

Bioreductions and biooxidations, although they can be highly selective, have often been limited by the requirement for expensive reducing or oxidizing biological cofactors. Wolfgang Kroutil of the University of Graz reports (*J. Org. Chem.* **2003**, *68*, 402-406. <u>DOI</u>) that aqueous suspensions of the whole lyophilized cells of *Rhodococcus ruber* DSM 44541 show alcohol dehydrogenase activity even in the presence of high concentrations of isopropanol or acetone. The organic co-solvent then serves as the "co-factor", driving reduction or oxidation. At the end of the reaction, the mixture is centrifuged, and the organic solvent is dried and concentrated. This promises to be an easily scalable preparative method.

The usual selectivities are observed, with aryl alkyl ketones and alkyl methyl ketones being reduced with high enantioselectivity (**1** -> **2** and **3** -> **4**)). That **5** is reduced to **6** with high ee, with the reducing enzymes differentiating between an ethyl and an *n*-pentyl group, is even more impressive.

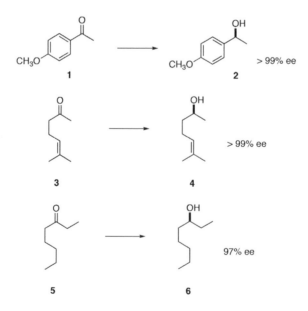

The 2-tetralone **7** (R = R' = H) is reduced to the alcohol **8** with respectable enantioselectivity. An intriguing question is, what would happen with R or R' = alkyl? Would one enantiomer reduce more rapidly than the other, perhaps with high diastereoselectivity? Could the other enantiomer (especially R = alkyl) epimerize under the reaction conditions?

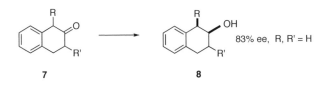

83% ee, R, R' = H

7   8

The selective reduction of **5** suggests that **9** and/or **12** might reduce with high enantioselectivity. This would open an inexpensive route to enantiomerically-pure epoxides, important intermediates for organic synthesis.

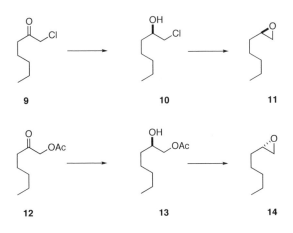

9   **10**   **11**

12   **13**   **14**

3

Catalytic Enantioselective Synthesis

January 26, 2004

The saga of efficient enantioselective catalysis by the amino acid proline continues. Nearly simultaneously, Dave MacMillan of Caltech and Yujiro Hayashi of the Tokyo University of Science reported (J. Am. Chem. Soc. 125: 10808, 2003; Tetrahedron Lett. 44: 8293, 2003) that exposure of an aldehyde **1** or ketone **4** to nitrosobenzene and catalytic proline gives the oxamination products **2** and **5** in excellent yield and ee. Reduction of **2** is reported to give the terminal diol **3** in 98% ee. The N-O bond can also be reduced with $CuSO_4$. The importance of prompt publication is underlined by these two publications – the MacMillan paper was submitted in July, and the Hayashi paper in August.

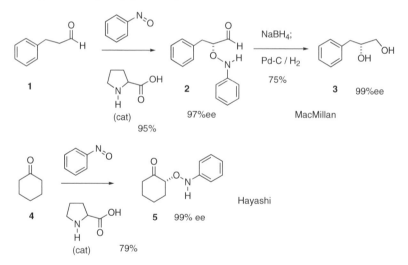

Many methods have been developed for asymmetric allylation. One of the best is the procedure reported by Masahisa Nakada of Waseda University (J. Am. Chem. Soc. 125: 1140, 2003). This uses the inexpensive allyl or methallyl chlorides directly. The reduction of the chloride with Mn metal is catalyzed by $CrCl_2$. When the addition is carried out in the presence of 10 mol % of the enantiomerically-pure ligand **7**, the product is formed in high yield and ee.

4

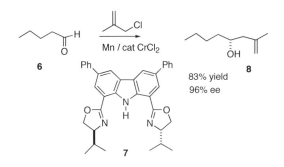

6

7

8

83% yield
96% ee

One of the severest challenges of asymmetric synthesis is the direct enantioselective construction of quaternary stereogenic centers. Brian Pagenkopof of the University of Texas has reported (Chem. Communications 2003: 2592) that alkynyl aluminum reagents will open a trisubstituted epoxide such as **10** at the more substituted center, with inversion of absolute configuration. As the epoxide **10** is available in high ee from **9** by the method of Yian Shi of Colorado State (J. Am. Chem. Soc. 119: 11224, 1997), this opens a direct route to quaternary cyclic stereogenic centers.

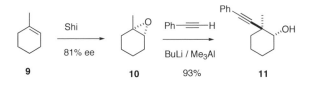

9   81% ee   10   93%   11

Enantioselective Synthesis of Borrellidin

February 2, 2004

There are two criteria for judging any total synthesis: the importance of the molecule that has been prepared, and the creativity evidenced in the synthetic route. When the natural product has only two rings, as with borrelidin **1**, the standards are even higher. The enantioselective total synthesis of borrelidin by Jim Morken of the University of North Carolina (J. Am. Chem. Soc. 125: 1458, 2003) more than exceeds those standards.

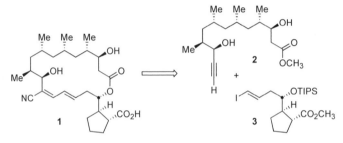

Borrelidin **1** has attracted attention because it inhibits angiogenesis, and so potentially blocks tumor growth, with an IC_{50} of 0.8 nM. Retrosynthetic analysis of **1** led the investigators to the prospective intermediates **2** and **3**. To assemble these two fragments, they interatively deployed the elegant enantio- and diastereoselective intermolecular reductive ester aldol condensation that they had recently developed. This transformation is exemplified by the homologation of **4** to **6** catalyzed by the enantiomerically-pure Ir complex **5**.

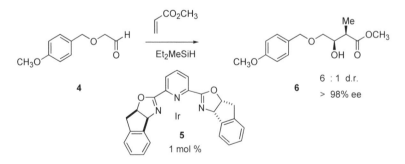

The final stages of the synthesis illustrate both the power and the current limitations of transition-metal mediated C-C bond formation. Coupling of **2** and **3** led to the ene-yne **7**. Pd-mediated hydrostannylation of the alkyne proceeded with high geometric control, but tended to

6

give the undesired regioisomer. The authors found that with the acetate, the ratio could be improved to 1 : 1. Iodination followed by Stille coupling then gave the dienyl nitrile **9**.

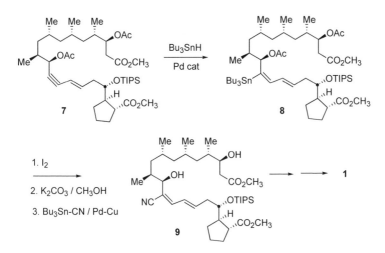

Enantioselective Ring Construction

February 9, 2004

New methods are being developed for the enantioselective construction of both heterocyclic and carbocyclic rings. Justin DuBois of Stanford reports (J. Am. Chem. Soc. 125: 2029, 2003) that his intramolecular Rh-mediated nitrene C-H insertion cyclizes **1** to **2** with high diastereoselectivity. The N,O-acetal opens smoothly with the alkynyl zinc, again with high diastereocontrol. The tosylate is stable to the alkyne addition conditions, but after reduction of the alkyne to the alkene the tosylate is readily displaced, to give **4**. After osmylation, the sulfamate undergoes smooth S$_N$2 displacement with cyanide ion, leading, after reduction, to the indolizidine **6**.

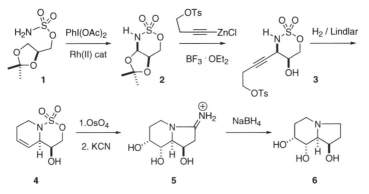

Sundarababu Baskaran of IIT-Madras offers (Organic Lett. 5: 583, 2003) an alternative route to indolizidines. Exposure of the epoxide **7** to Lewis acid followed by reduction leads to **11** as a single diastereomer. The authors hypothesize that this rearrangement is proceeding via intermediates **8** - **10**. Tosylation of **11** followed by homologation leads to the Dendrobatid alkaloid **12**.

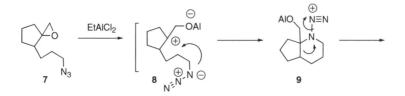

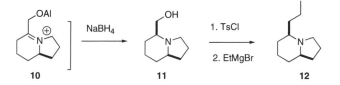

Intramolecular carbon-carbon formation to convert inexpensive, enantiomerically-pure carbohydrates directly to highly functionalized, enantiomerically-pure carbocycles has long been a goal of organic synthesis. Ramón J. Estévez of the University of Santiago in Spain reports (Organic Lett. 5:1423, 2003) that TBAF smoothly converts the triflate derived from **13** into **14**, without competing β-elimination.

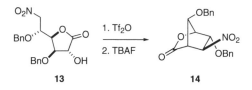

New Routes to Heterocycles

February 16, 2004

The efficient construction of substituted heterocycles is central to medicinal chemistry. Yoshinori Kondo of Tohoku University reports (J. Am. Chem. Soc. 125: 8082, 2003) that the novel base **2** will, in the presence of ZnI$_2$ and an aldehyde, deprotonate heterocycles, to give the hydroxyalkylated products **3** and **5**. Benzene rings also participate – **6** is converted to **7**.

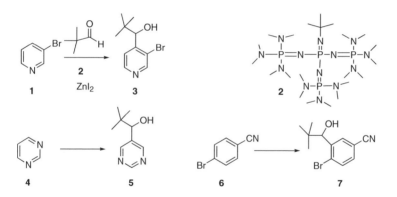

Free radical cyclizations have often been carried out with tin reagents, which are toxic, and an environmenal hazard. As an alternative, phosphorus reagents have been developed, but these suffer from the shortcoming that they are aqueous, and so it is difficult to get them into contact with the organic substrate. John A. Murphy of the University of Strathclyde in Glasgow reports (Organic Lett. 5: 2971, 2003) that diethyl phosphine oxide (DEPO), which is soluble in organic solvents as well as in water, serves efficiently as a hydride source and radical mediator, smoothly cyclizing **8** to the indolone **9**. The oxidized phosphonic acid byproduct is easily separated by aqueous extraction.

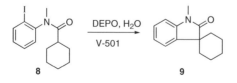

The Wittig reaction efficiently olefinates aldehydes and ketones, but not esters or amides. Several early-transition-metal approaches have been taken to this problem. Recently, Takeshi Takeda of the Tokyo University of Agriculture and Technology reported (Tetrahedron Lett. 44: 5571, 2003) that the titanocene reagent can effect the condensation of an amide **10** with a thioacetal **11** to give the enamine **12**. On hydrolysis, **12** is converted into the ketone **13**. When the reaction is intramolecular, reduction proceeds all the way, to give the pyrrolidine **15**.

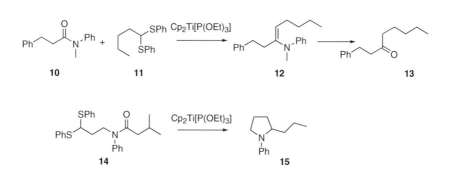

Total Synthesis of the Galbulimima Alkaloid GB 13

February 23, 2004

Lew Mander of the Australian National University recently reported (*J. Am. Chem. Soc.* **2003**, *125*, 2400) the total synthesis of the pentacyclic alkaloid GB 13 **3**, which had been isolated from the bark of the rain forest tree *Galbulimima belgraveana*. In the course of the synthesis, he took full advantage of benzene precursors, while at the same time carefully establishing each of the eight stereogenic centers of **3**.

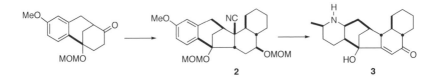

The core tricyclic ketone **1** was assembled by Birch reduction of 2,5-dimethoxybenzoic acid, followed by alkylation with 3-methoxybenzyl bromide, to give **4**. Acid-catalyzed electrophilic cyclization of **4** gave the tricyclic ketone **5**, which on decarboxylation and protection gave **1**.

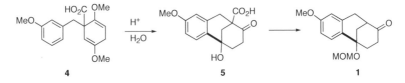

Diazo transfer to **1** followed by irradiation in the presence of bis-(trimethylsilyl)amide led to ring contraction with concomitant carbonyl extrusion, to give **7**. Dehydration to the nitrile followed by selenation then set the stage for a highly diastereoselective ytterbium-catalyzed Diels-Alder reaction, to give, after reduction and protection, the pentacyclic intermediate **2**.

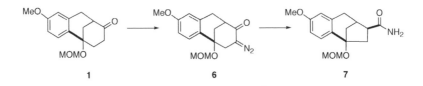

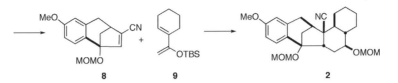

8   **9**   **2**

Intermediate **2** appears to have *two* extraneous carbons, the nitrile, *and* one of the carbons of the aromatic ring. In fact, the aromatic carbon was carried all the way through, to appear as the α-methyl group on the piperidine ring. Birch reduction of **2** deleted the superfluous nitrile and reduced the aromatic ring, to give, after hydrolysis, the enone **10**. Eschenmoser fragmentation of the intermediate epoxy ketone then gave the keto alkyne **11**. The subsequent condensation with hydroxylamine followed by reduction proceeded with spectacular (but anticipated) stereocontrol, to establish the three stereogenic centers of the trisubstituted piperidine ring. Oxidation of **12** then gave the enone **3**.

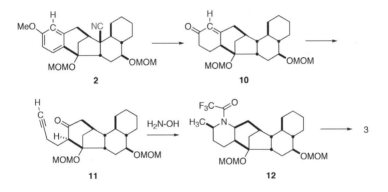

2   **10**

11   **12**   3

13

Total Synthesis of Ingenol

March 1, 2004

The total synthesis of the tetracyclic *Euphorbia* tetraol ingenol **3** reported by Keiji Tanino of Hokkaido University (J. Am. Chem. Soc. 125: 1498, 2003) illustrates the power of diastereoselective carbocationic rearrangements, as exemplified by the conversion of **1** to **2**.

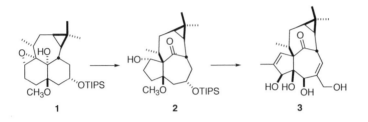

The construction of the tricyclic epoxide depended on several highly diastereoselective transformations. The addition of lithio t-butyl acetate to ketone **4** proceeded to give **5** as a single diastereomer, even though the ketone is flanked by a quaternary center. The authors speculate that lithium chelation with the methyl ether directed addition. Even more spectacular was the cyclization of the propargylic acetate **7** to **10**. The Co complex activated the acetate for ionization, while at the same time establishing the proper geometric relationship for bond formation. Dissolving metal reduction of the Co complex then gave the alkene.

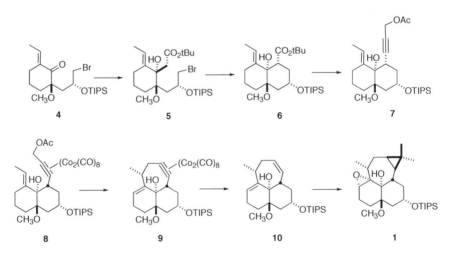

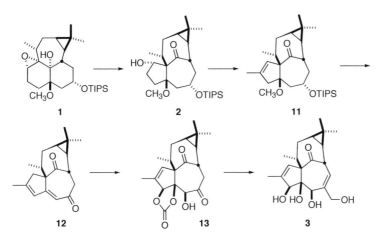

1   2   11

12   13   3

The elegant pinacol rearrangement of **1** to **2**, mediated by $(ArO)_2AlCH_3$, exposed a ketone that might usually need to be protected.   In this case, however, the ketone is so buried in the inside-outside ingenol skeleton that it is unreactive. After several further manipulations, a spectacular osmylation of the diene **12** led to ingenol **3**, in an overall 45-step sequence.

The ingenol **3** prepared by this route was racemic. It is interesting to speculate how one might efficiently prepare **4** or its precursors in enantiomerically-pure form.

15

Best Synthetic Metods

March 8, 2004

    Benzyl ethers are among the most commonly used protecting groups for alcohols. Usually, these are prepared using an excess of NaH and benzyl bromide. Okan Sirkecioglu of Istanbul Technical University has found (Tetrahedron Lett. 44: 8483, 2003) that heating a primary or secondary alcohol neat with benzyl chloride and a catalytic amount of Cu(acac)$_2$ smoothly yields the benzyl ether, with evolution of HCl. The reaction can also be run in solvent THF, but it proceeds more slowly. Note that primary alcohols react more quickly than secondary alcohols.

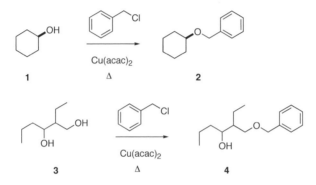

    Usually, one would consider the conversion of an aldehyde or a ketone to the ether to be a two-step process. There are, however, catalysts that will effect this conversion in a single step, as exemplified by the work of Shang-Cheng Hung of Academia Sinica in Taipei (Tetrahedron Lett. 44: 7837, 2003). Although one might choose to use this as another alternative for preparing benzyl ethers, note that it also offers an efficient procedure for preparing unsymmetrical *dialkyl* ethers, which can sometimes be a difficult task. The diastereoselectivity of the process is particularly impressive.

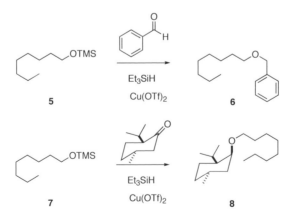

Several years ago, Jim-Min Fang of National Taiwan University reported that an aldehyde **9** was smoothly converted into the corresponding nitrile **10** by iodine in aqueous ammonia. He has now observed (J. Org. Chem. 68: 1158, 2003) that the intermediate nitrile can be carried on in situ to the amide **11**, the tetrazole **12**, or the triazine **13**.

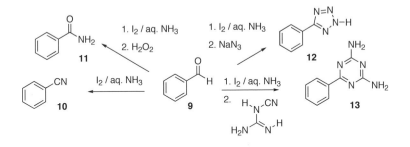

New Methods for Carbon-Carbon Bond Formation

March 15, 2004

The construction of carbon-carbon bonds is fundamental to organic synthesis. Recently, three new methods have been reported, each of which has substantial potential for the synthesis of highly functionalized target molecules.

Shmaryahu Hoz of Bar-Ilan University reports (J. Org. Chem. 68: 4388, 2003) that alkyl boranes couple with dinitro aromatic rings such as **1** to give the alkylated aromatic, with loss of one of the nitro groups. This reaction shows remarkable regioselectivity, as illustrated by the formation of **2**. Much more complex alkyl boranes participate also, as illustrated by the coupling of the 9-BBN derivative **3**. The reaction proceeded to give **4** as a single diastereomer.

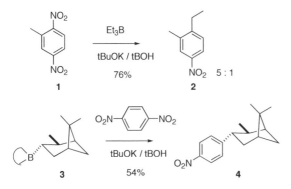

The Heck reaction is well developed as a method for alkylating aromatics. Frank Glorius of the Max-Planck-Institute, Mülheim, reports (Tetrahedron Lett. 44: 5751, 2003) that chloroacetamides and bromoacetonitrile can also be activated by catalytic Pd to give the coupled products. This reaction works well with enol ethers, to give highly functionalized alkenes, but it also works well with a simple cyclic alkene.

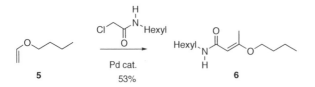

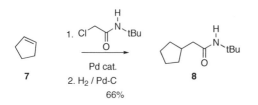

7 → 8

1. Cl—CH₂—C(=O)—N(H)—tBu
Pd cat.
2. H₂ / Pd-C
66%

The Pd-catalyzed homologation of aromatic rings does not necessarily require an aryl halide or other leaving group. The presence of a suitably disposed chelating group is sufficient to mediate metalation, followed by coupling. Masahiro Miura of Osaka University reports (J. Org. Chem. 68: 5236, 2003) that the tertiary alcohol **9** smoothly undergoes ortho palladation and coupling. The reaction can be limited to monocoupling, or extended to double homologation, to give **10**. Interestingly, the tertiary alcohol itself can also serve, with loss of acetone, as a precursor to the aryl Pd species.

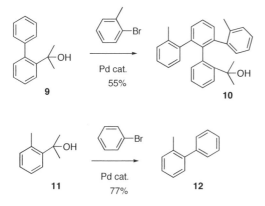

9

Pd cat.
55%

10

11

Pd cat.
77%

12

Mini-Review: Organic Reactions in Ionic Liquids

March 22, 2004

Ionic liquids are organic salts that are liquid at or near room temperature. It has been found recently that such liquids can be useful solvents for organic reactions. Often, the organic products can be removed from the ionic liquid by extraction with, e.g., ether, without resorting to an aqueous workup. This can be particularly useful when a precious metal catalyst is used in the reaction. The catalyst often remains in the ionic liquid, so that the catalyst solution can be directly reused.

Roberta Bernini of the Univ. of Tuscia in Viterbo illustrated (Tetrahedron Lett. 44: 8991, 2003) the power of this approach with the $MeReO_3$-catalyzed oxidation of **1** to **2** in [bmim]BF_4. The product could be extracted with ether and the catalyst-containing ionic liquid recharged with substrate and H_2O_2 for four cycles before the conversion and yield started to drop off. This drop off may be due to the accumulation of water in the ionic liquid, which could be removed by distillation.

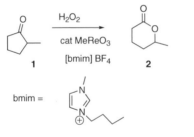

Alternative purification protocols are available. Zhaolin Sun of Lanzhou University reports (Tetrahedron Lett. 45: 2681, 2004) that the ionic liquid TISC was specifically designed to promote Beckmann rearrangement. TISC is not soluble in water, so the product caprolactam was easily removed from the ionic liquid by extraction with water.

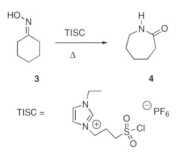

The counterion of the ionic liquid can be tuned to achieve one desired reactivity or another. Martyn Earle of Queen's University, Belfast has observed (Organic Lett. 6: 707 , 2004) that the reaction of toluene can be directed toward any of the three products **6**, **7**, or **8**, depending on the ionic liquid used.

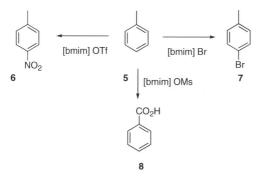

Ionic liquids have been used for many other reactions. Examples include Friedel-Crafts acylation (Tetrahedron Lett. 43: 5793, 2002), osmylation (Tetrahedron Lett. 43: 6849, 2002), Heck coupling (Tetrahedron Lett. 44: 8395, 2003), and Henry reaction (Tetrahedron Lett. 45: 2699, 2004).

Adventures in Polycyclic Ring Construction

March 29, 2004

The development of efficient strategies for the construction of designed polycyclic systems is one of the more challenging and intriguing activities of organic synthesis. The three approaches outlined here each depend on a planned sequence of events.

The intramolecular Diels-Alder reaction has long been a powerful method for polycyclic ring construction. Al Padwa of Emory University reports (J. Org. Chem. 68: 227, 12003) that on Rh catalysis, a diazoalkyne such as **1** is smoothly converted into the reactive furan **2**. Cyclization of **2** leads via **3** to the angularly-arylated product **4**.

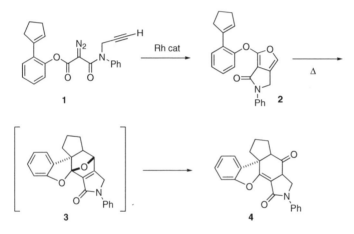

Scott Denmark of the University of Illinois reports (J. Org. Chem. 68: 8015, 2003) a *hetero* intramolecular Diel-Alder reaction of a nitro alkene **5**, followed by intramolecular dipolar cycloaddition of the resulting nitronate **6**, to give the tricycle **7**. Raney nickel reduction effected cleavage of the N-O bonds and reductive amination of the liberated aldehyde, to give, after acetylation, the angularly substituted cis-decalin **8**.

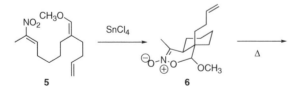

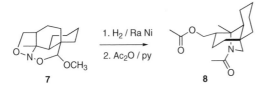

7   8

Sam Zard of the Ecole Polytechnique in Paliseau has developed elegant and affordable free-radical methods for C-C bond construction. In the context of the total synthesis of pleuromutilin, he recently reported (Organic Lett. 5: 325, 2003) that the free radical cyclization of **12** proceeded smoothly to give the eight-membered ring product **13**. The ketone **12** is easily prepared from m-toluic acid. It is a tribute to the efficacy of the cyclization procedure that the conformation drawn, the conformation required for cyclization, is the *less* stable chair available to **12**.

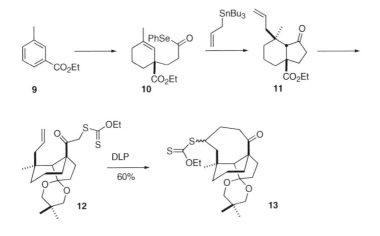

Synthesis of the Mesotricyclic Diterpenoids Jatrophatrione and Citlalitrione

April 5, 2004

Leo Paquette of Ohio State recently reported (J. Am. Chem. Soc. 125: 1567, 2003) the total synthesis of jatrophatrione **1** and citlalitrione **2**. These diterpenes, which share a central highly-subsituted 5-9-5 core, show remarkable tumor-inhibitory activity.

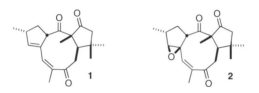

The 5-9-5 skeleton was assembled by the addition of the alkenyl cerate derived from **6** with the ketone **4**, to give **7**. Oxy-Cope rearrangement then gave the 5-9-5 enolate, which was quenched with methyl iodide to give **8**. The ketone **8** underewent spontaneous intramolecular ene cyclization, to give **9**.

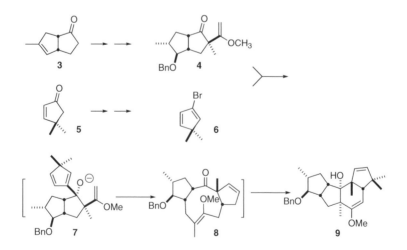

The transient 5-9-5 ketone **8** has two cis-fused rings. To invert the ring stereochemistry, the alkene **9** was oxidized to the enone **10**. After some experimentation, it was found that a CuH preparation would reduce the enone to give predominantly the trans-fused ketone. Monomesylation of the derived diol set the stage for Grob fragmentation to reopen the nine-membered ring, providing, after reduction, the alcohol **12**.

At this point, there were two problems in selective alkene functionalization to be addressed. Although all attempts at oxidation of the cyclopentene failed, intramolecular hydrosilylation proceeded smoothly, to give **13**. On exposure of the derived cyclic carbonate to Hg(O$_2$CCF$_3$)$_2$, the cyclononene then underwent allylic oxidation, to give **14**.

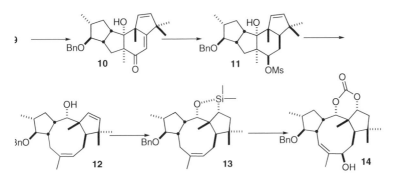

Attempts to functionalize the homoallylic alcohol **15** quickly revealed that this product of an intramolecular aldol condensation was sensitive to base. Fortunately, heating with thiocarbonyldiimidazole effected clean dehydration to give predominantly the desired regioisomer of the diene. Methanolysis followed by oxidation then gave the triketone **1**, which on epoxidation with MCPBA gave **2** as the minor component of a 3:1 mixture.

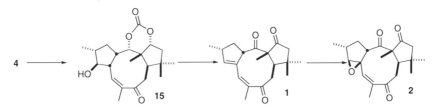

Best Synthetic Methods

April 12, 2004

The Strecker synthesis is the one-carbon homologation of an aldehyde to the α-amino nitrile. Robert Cunico of Northern Illinois University in DeKalb reports (Tetrahedron Lett. 44: 8025, 2003) a modified Strecker leading directly to the amide of the α-amino acid.

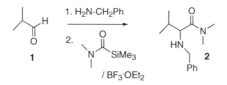

The conversion of a ketone to the halide is usually a two-step process. Akio Baba of Osaka University reports (J. Am. Chem. Soc. 124: 13690, 2002), the one-step reduction of ketones and aldehydes to the corresponding chlorides and iodides. It is noteworthy that the reaction proceeds even with aliphatic ketones.

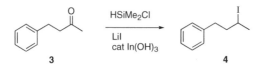

Convenient new methods for the preparative scale oxidation of alcohols to ketones and carboxylic acids are always welcome. T. Punniyamurthy of the Indian Institute of Technology Guwahati reports (Tetrahedron Lett. 44: 6033, 2003) that 30% aqueous H_2O_2 catalyzed by a Co salen complex effects this transformation.

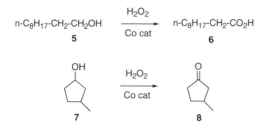

Usually, one would expect that an acrylate ester would be prepared by the acylation of an alcohol with acryloyl chloride. Jonathan M.J. Williams of the University of Bath reports (Tetrahedron Lett. 44: 5523, 2003) that this acylation can also be effected with the mild combination of Ph_3P and maleic anhydride. The acrylate esters so prepared are interesting as polymerization precursors, and as Diels-Alder dienophiles. The allylic acrylates invite tandem conjugate addition / Ireland Claisen rearrangement.

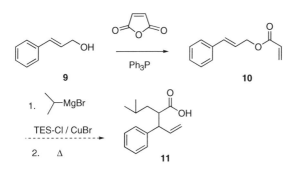

The Grubbs Reaction in Organic Synthesis

April 19, 2004

Alkene metathesis (e.g. **1 + 2 → 3**) has been known at least since the 1950's. Until Robert Grubbs of Caltech developed stable and versatile Ru catalysts for this transformation, however, this reaction was little used.

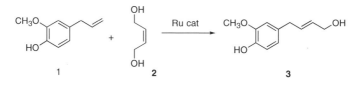

Professor Grubbs recently published (J. Am. Chem. Soc. 125: 10103, 2003; J. Am. Chem. Soc. 125: 11360, 2003) two detailed articles on activation and selectivity in this reaction. The first article addresses variations on catalyst design. The second paper defines several types of alkenes, and lays out rules that allow one to predict which pairs of alkenes will dimerize efficiently. While it is not possible to summarize all of their results in this limited space, some highlights include:

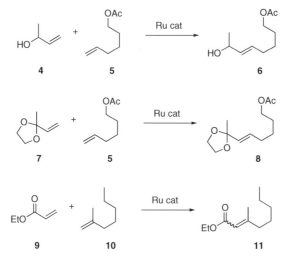

It is also not possible to even partially review all the many applications in synthesis that have already been demonstrated for the Grubbs reaction. I have chosen to focus on three papers. For those desiring to deploy the Grubbs catalyst only from time to time, storage and handling become serious issues. We have found (J. Org. Chem. 68: 6047, 2003) that the commercial catalyst dissolved in paraffin wax can be stored exposed to the laboratory atmosphere for many months and still retain full activity. The example above of **1 + 2 → 3** is taken from that paper.

Steve Martin of UT Austin has reported (J. Org. Chem. 68: 8867, 2003) a detailed study of the synthesis of bridged azabicyclic structures via ring-closing alkene metathesis. Some examples of his work include the conversion of **12** to **13**, efficiently forming six, seven and eight membered rings. He also demonstrated five-membered ring formation with the conversion of **14** to **15**, which has the cocaine skeleton.

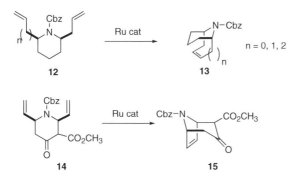

A real concern when attempting the Grubbs reaction with a complex substrate is the stability of other alkenes. J. Alberto Marco of the University of Valencia, Spain has shown (J. Org. Chem. 68: 5672, 2003) that **16** is converted to **17** without isomerization of the Z alkene. Less congested alkenes might not be so resistant.

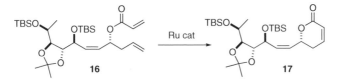

C-N Ring-forming Reactions by Transition Metal-catalyzed Intramolecular Alkene Hydroamination

April 26, 2004

Alkene hydroamination has been known for many years, but has been little used as a method in organic synthesis. Tobin Marks of Northwestern recently published a series of three papers that will make this transformation much more readily accessible. In the first (J. Am. Chem. Soc. 125: 12584, 2003) he describes the use of a family of lanthanide-derived catalysts for intermolecular hydroamination of alkynes (to make imines, not illustrated) and alkenes. With aliphatic amines, the branched (Markownikov) product is observed, **1 → 2**. With styrenes, the linear product is formed. When two alkenes are present, the reaction can proceed (**3 → 4**) to form a ring, with impressive regioselectivity.

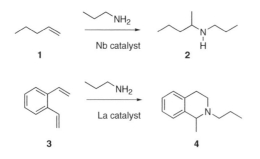

The products from alkene hydroamination are inherently lightly functionalized. To address this possible deficiency, Professor Marks also reported (J. Am. Chem. Soc. 125: 15878, 2003) the cyclization of amino dienes such as **5**. The cyclizations proceed with high selectivity for the cis-2,6-dialkyl piperidines, and with a little lower selectivity for the trans 2,5-dialkyl pyrrolidine. The product alkenes are ~95% E, the balance being a little Z alkene and the terminal alkene.

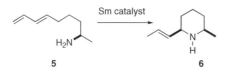

More complex substrates can also be cyclized efficiently using these catalysts. Professor Marks and his colleague Frank McDonald, now at Emory University, report (J. Org. Chem. 69, 1038, 2004) that the amino diene **7** cyclizes to **8** with 81:19 diastereoselectivity. It is particularly noteworthy that with each of these ring-forming reactions, the free amines are employed, avoiding inefficiencies of protection and of protecting group removal.

7   8

Synthesis of (+)-Phomactin A

May 3, 2004

The diterpene **(+)**-Phomactin A **4** is an antagonist of platelet activating factor. The preparation of **4** recently reported (J. Am. Chem. Soc. 125: 1712, 2003) by Randall Halcomb of the University of Colorado elegantly illustrates the use of readily-available natural products as starting materials for natural product synthesis.

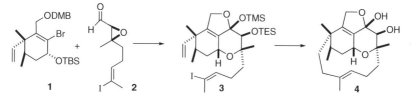

The synthetic plan called for a late-stage intramolecular reductive coupling of the iododiene **3** to establish the macrocyclic ring of **4**. The iododiene **3** was to be assembled by condensation of the highly-substituted cyclohexene **1** with the aldehyde **2**.

The aldehyde **2** was prepared from the inexpensive geraniol ether **5**. Selective ozonolysis followed by Wittig homologation gave the bromodiene, which was converted via dehydrobromination and alkylation to the alkyne **6**. Regioselective hydridozirconation followed by iodination of the C-Zr bond gave the alkenyl iodide **7** with high geometric control. The two stereogenic centers of **2** were then established by Sharpless asymmetric epoxidation.

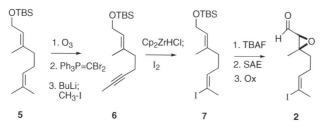

The preparation of the cyclohexene **1** began with pulegone **8**, available commercially in high enantiomeric purity. Methylation followed by retro aldol condensation to remove the unwanted isopropylidene group gave 2,3-dimethylcyclohexanone, which on bromination-dehydrobromination gave **9**. Vinylation followed by alkylative enone transposition gave **11**, which was brominated over several steps to give **12**. Conditions to reduce the ketone **11** directly to the axial alcohol were unavailing, so the dominant pseudoequatorial alcohol from NaBH$_4$ reduction was inverted, to give **1**.

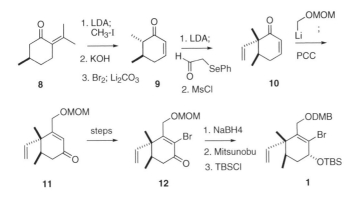

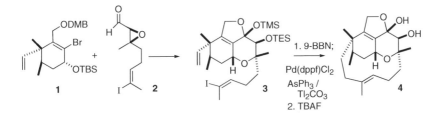

Condensation of **1** with **2** led to **3**, setting the stage for the key macro ring closure. Happily, conditions could be developed to effect this important transformation, a B-alkyl Suzuki coupling. The ligand dppf is 1,1'-bisdiphenylphosphinoferrocene. The use of AsPh₃, rather than a phosphine, as the supporting ligand was important, as was the use of the thallium base.

Enzymes in Organic Synthesis

May 10, 2004

Enzymes have many applications in organic synthesis. One of the most common is the resolution of a secondary alcohol. A shortcoming of this approach is that the separation of the residual alcohol from the product ester has required column chromatography, adding to the expense. Louisa Aribi-Zouioueche of the University of Annaba, Algeria, and Jean-Calude Fiaud of the University Paris-Sud, Orsay, have been exploring (Tet. Lett. 45: 627, 2004) the use of succinic anhydride as an alternative to the more typical vinyl acetate or isopropenyl acetate acylating agents. Using this procedure, the residual alcohol and the product ester can be separated by simple acid-base extraction. The key question they had to address was whether or not the enantioselectivity of the acylation was maintained. The results were mixed, but in at least one case, the quinoline alcohol **1**, the enantioselectivity was *improved* using this protocol.

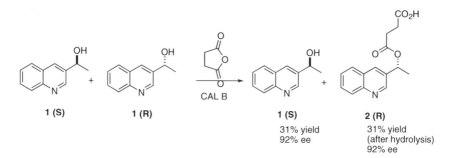

1 (S)	1 (R)		1 (S)	2 (R)
		CAL B	31% yield 92% ee	31% yield (after hydrolysis) 92% ee

Enzymes can also be used to reduce organic substrates, as illustrated by the conversion of cyclohexanone **3** to cyclohexanol **5**. A shortcoming of this approach is that a stoichiometric amount of the reducing cofactor, in this case NADH, is required. This need can be met by simultaneously oxidizing a sacrificial substrate, so as to regenerate the NADH. Ikuo Uedo of Osaka University has developed (J. Org. Chem. 67: 3499, 2002) an interesting alternative. The inexpensive Hantzsch carboxylate will regenerate NADH from NAD^+. Using this procedure, they were able to observe up to ten turnovers of the cofactor.

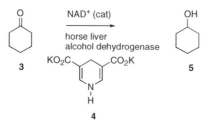

It is interesting that NADH is also required as a stoichiometric co-factor in enzymatic oxygenation processes. In a detailed study of styrene monooxygenase (StyA), Andreas Schmid of the ETH/Zurich showed (J. Am. Chem. Soc. 125: 8209, 2003) that Cp*Rh(bpy)(H$_2$O)$^{2+}$ in combination with sodium formate served effectively to regenerate the NADH. Using this combination, epoxidation of aryl alkenes such as **6**, **8** and **10** proceeded in high enantiomeric excess.

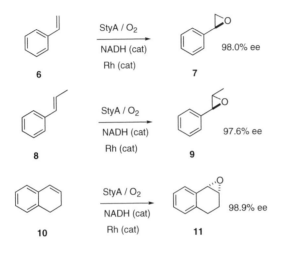

Adventures in Polycarbocyclic Construction

May 17, 2004

As the computational methods used in pharmaceutical development have improved, receptor binding analysis has led to many potential new drug candidates that are polycyclic. Such leads are often not pursued, however, because of the perception that even if it turned out to be active, an enantiomerically-pure polycyclic agent would be too expensive to manufacture. Taking this as a challenge, academic research groups continue to develop clever approaches for the efficient synthesis of complex polycarbocyclic target structures. Three recent approaches are outlined here.

Hee-Yoon Lee of the Korea Advanced Institute of Science & Technology (KAIST) in Daejon reported (J. Am. Chem. Soc. 125: 10156, 2003) that on heating, the imine **1** is cleanly converted into the tricyclic **2**. The reaction presumably proceeds via insertion of the alkylidene carbene **3** into the alkene, to give the unstable alkylidene cyclopropane **4**. The authors suggest that **4** opens to the diradical **5**, which then cyclizes. It is striking that the geometry of the starting alkene **1** is retained in the product **2**. It is possible that in fact there is a concerted pathway for the opening of **4** and simultaneous insertion, to give **2**.

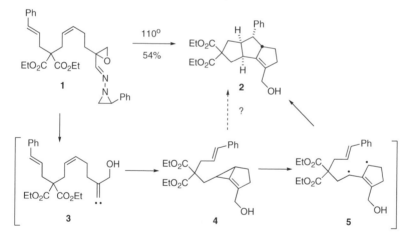

Hiroto Nagaoka of Tokyo University of Pharmacy and Life Science has reported (Tetrahedron Lett. 44: 4649, 2003) the tandem reduction – Dieckmann cyclization of the esters **6** and **9**. It is striking that the geometry of the starting alkene dictates the ring fusion of the product. Both 5/5 and 6/5 systems can be prepared this way.

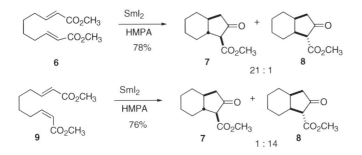

6 → **7** + **8**

SmI₂ / HMPA / 78%

21 : 1

9 → **7** + **8**

SmI₂ / HMPA / 76%

1 : 14

The furanoterpene 15-acetoxytubipofuran **12** shows cytotoxicity against B-16 melanoma cells. E. Peter Kündig of the University of Geneva has reported (J. Am. Chem. Soc. 125: 5642, 2003) a concise asymmetric synthesis of **12**, based on the addition of lithio ethyl vinyl ether to the chromium tricarbonyl-activated benzaldehyde **10**. In the course of the organometallic addition, five carbon-carbon bonds are formed.

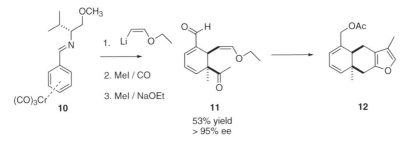

10

1. Li ⌁O⌁
2. MeI / CO
3. MeI / NaOEt

11
53% yield
> 95% ee

12

Construction of Enantiometrically-Pure Heterocycles

May 24, 2004

As most pharmaceuticals are heterocyclic, there is continuing interest in methods for the direct enantioselective construction of heterocycles. Greg Fu of MIT reports (J. Am. Chem. Soc. 125: 10778, 2003) that the addition of the dipole **1** to alkynes is catalyzed by CuI, and that in the presence of the planar–chiral ligand **2** the reaction proceeds in high enantiomeric excess. The ee is maintained with aryl-substituted alkynes, and is higher when there are alkyl substituents on the heterocyclic ring of **1**.

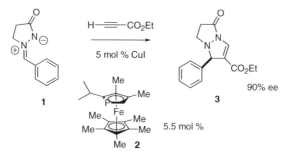

Many methods have been developed for the enantioselective synthesis of unnatural α-amino acids. Jeff Johnston of Indiana University reports (J. Am. Chem. Soc. 125: 163, 2003) coupling the asymmetric alkylation of O'Donnell with intramolecular radical cyclization, leading to what appears to be a general method for the enantioselective construction of indolines.

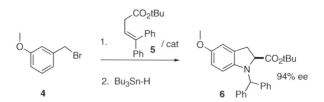

Takeo Kawabata of the Institute for Chemical Research associated with Kyoto University reports (J. Am. Chem. Soc. 125: 13012, 2003) that unnatural amino acids can also be used to assemble four-, five-, six-, and seven-membered cyclic amines having *quaternary* stereogenic centers. Given the conventional wisdom that ester enolates are sp^2-hybridized, this memory effect is remarkable.

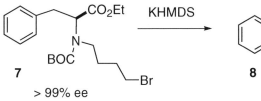

KHMDS

7

> 99% ee

8

97% ee

Best Synthetic Methods

May 31, 2004

From time to time, synthetic transformations are reported that appear to be particularly convenient. One such is the procedure described by Paul Hanson of the University of Kansas (Tetrahedron Lett. 44: 7187, 2003) for the conversion of an alcohol to the amine. The imine **2**, easily prepared by the addition of maleimide to furan, couples under Mitsubobu conditions with the **1** to give the imide **3**, contaminated with the usual impurities from the condensation. The crude imide **3** smoothly polymerizes under Ru metathesis conditions to give polymeric **3**, from which the impurities are easily washed away. Exposure of the washed polymer to hydrazine then liberates the pure free amine **4**.

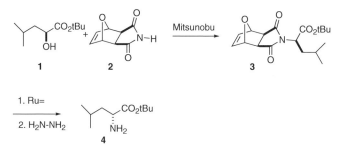

Benzyl and substituted benzyl protecting groups are ubiquitous in organic synthesis. For base-sensitive substrates, the benzyl imidate, e.g. **6**, is often used to install this group. Amit Basu of Brown University reports (Tetrahedron Lett. 44: 2267, 2003) that in situations such that the usual acidic promoters do not work, metal triflates can be effective. Lanthanum triflate worked particularly well, giving both high yield and high conversion in five minutes at room temperature.

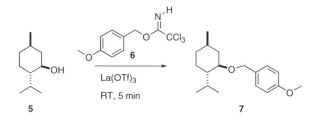

The oxidation of an alcohol to the aldehyde or ketone on large scale would ideally be carried out with an inexpensive, easily handled reagent at ambient temperature. Scott Hoerrner of Merck Process, Rahway, NJ reports (Organic Lett. 5: 285, 2003) that stoichiometric iodine in the presence of catalytic TEMPO cleanly converts the sensitive alcohol **8** to the aldehyde **9**. While this method was developed particularly for easily oxidized heteroaromatics such as **9**, in fact the procedure works well for ordinary alcohol to aldehyde and alcohol to ketone oxidations.

There may be circumstances such that oxidation of an alcohol to the ketone with Cr is warranted. Mark Trudell of the University of New Orleans reports (Tetrahedron Lett. 44: 2553, 2003) that such oxidations can be carried out with catalytic Cr and stoichiometric periodic acid, as illustrated by the conversion of **10** to **11**.

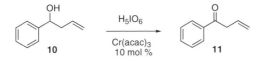

Synthesis of (+)-4,5-Deoxyneodolabelline

June7, 2004

The dolabellanes, represented by 3-hydroxydolabella-4(16), 7, 11(12)-triene-3,13-dione **1** and the neodolabellanes, represented by (+)-4,5-deoxyneodolabelline **2**, are isolated from both terrestrial and marine sources. They show cytotoxic, antibiotic and antiviral activity. The recent synthesis of (+)-4,5-deoxyneodolabelline **2** by David Williams of Indiana University (J. Am. Chem. Soc. 125: 1843, 2003) highlights both the strengths and the challenges of the current state of the art in asymmetric synthesis.

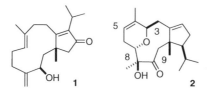

The synthetic plan was to assemble both the dihydropyran **3** and the cyclopentane **4** in enantiomerically-pure form, then to effect Lewis acid-mediated coupling of the allyl silane of **4** with the anomeric ether of **3** to form a new stereogenic center on the heterocyclic ring. A critical question was not just the efficiency of this step, but whether or not the desired stereocontrol could be achieved at C-3.

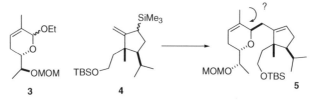

The construction of the heterocycle **3** started with enantiomerically-pure ethyl lactate. Protection, reduction and oxidation led to the known aldehyde **6**. Chelation-controlled allylation gave the monoprotected-diol **7**. Formation of the mixed acetal with methacrolein followed by intramolecular Grubbs condensation then gave **3**. The dihydropyran **3** so prepared was a 1:1 mixture at the anomeric center.

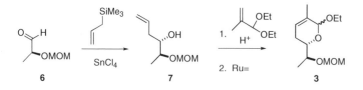

The preparation of the cyclopentane **4** proved to be more of a challenge. Rather than attempt an enantioselective synthesis, racemic **11** was prepared in straightforward fashion from commercially-available 2-methylcyclopentenone, by conjugate addition followed by alkylation of the regenerated ketone enolate. Ozonolysis followed by selective reduction then led to **11**. Resolution was accomplished by enantioselective reduction of the racemic ketone, to give a 1:1 mixture of separable diastereomers. Reoxidation of one of the diastereromers gave ketone **11**, which was determined to be a 96:4 mixture of enantiomers. Homologation followed by allylic silylation then gave **4** as an inconsequential mixture of diastereomers.

Condensation of the allyl silane **4** with **3** proceeded to give exclusively the desired *trans* dihydropyran **5**. McMurry coupling of the derived keto aldehyde gave the diol **13** as a mixture of diastereomers. Oxidation of the mixture gave **2** and its C-8 diastereomer in a ratio of 8:1.

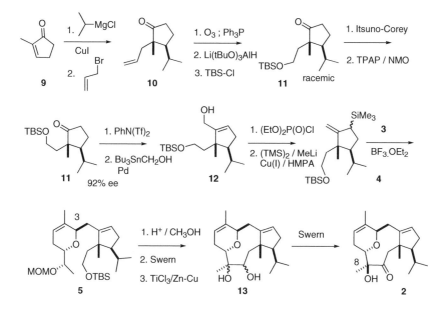

New Methods for Carbon-Carbon Bond Formation

June14, 2004

The development of new methods for carbon-carbon bond formation is at the heart of organic synthesis. The most desirable methods are those that are easily practiced at scale, operate near ambient temperature, and that do not require strong acid or base. David C. Forbes of the University of South Alabama and Michael C. Standen of Synthetech in Albany, OR report (Organic Lett. 5: 2283, 2003) that the crystalline salt **2**, which can be stored, smoothly converts aldehydes to epoxides, without any additional added base. The reaction is apparently proceeding by the loss of CO_2 from **2** to give the intermediate sulfonium methylide.

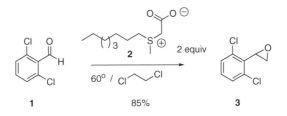

Fumie Sato of the Tokyo Institute of Technology has extensively developed applications of titanacyclopropenes such as **6**. He now reports (Organic Lett. 5: 67, 2003) the extension of this work to ynamides such as **4** and **10**. The titanacycle **6** derived from **4** can be protonated to give the cis alkene **7**. Titanacycle **6** also adds to an aldehyde, to give the geometrically-defined allylic alcohol **8**. Alternatively, the titanacycle prepared from another alkyne such as **9** will add to the ynamide **10**, to give the diene **11**. Titanium isopropoxide and 2-propylmagnesium chloride are inexpensive, and these couplings do not require catalysis by other transition metals.

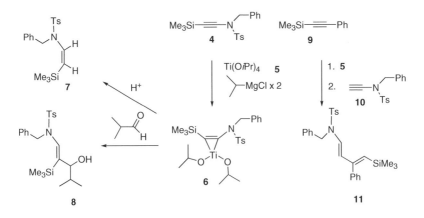

Usually, the construction of carbocyclic rings requires the preparation of highly functionalized intermediates. Youquan Deng of the Lanzhou Institute of Chemical Physics reports (Tetrahedron Lett. 48: 2191, 2003) that simple acid-mediated equilibration of 1-dodecene **12** gives a remarkably efficient conversion to cyclododecane **13**. The authors speculate that the peculiar thermodynamic stability of **13** favors this transformation.

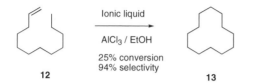

Strategies for Enantioselective Synthesis

June21, 2004

The direct enantioselective synthesis of quaternary centers is one of the enduring challenges of organic synthesis. Claude Spino of the Université de Sherbrooke reports (J. Am. Chem. Soc. 125: 12106, 2003) that organocuprate addition to the secondary pivalate **1** proceeds with outstanding diastereocontrol, to give **2** with an enantiodefined alkylated quaternary center. The homoallylic alcohols so prepared were efficiently converted into the corresponding enantiomerically-pure protected α, α-dialklyated α-amino acids, such as **3**.

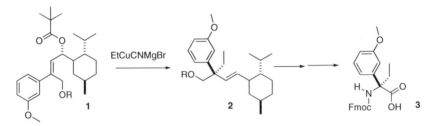

Tetsuaki Tanaka of Osaka University has reported (Tetrahedron Lett. 45: 75, 2004) what appears to be a general route to alkylated quaternary centers, based on the Ti-mediated addition of allylmagnesium chloride to the Sharpless-derived epoxy ether **4**. Remarkably, the conversion of **6** to **7** works equally well.

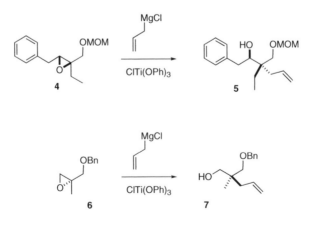

Enantiomerically-pure sulfoxides are readily available. Ilan Marek of Technion-Israel Institute of Technology reports (J. Am. Chem. Soc. 125: 11776, 2003) that alkyne-derived sulfoxides such as **8** can be used to direct the addition of an allylic organometallic, prepared *in situ*, to an aldehyde **9**. Both the secondary alcohol, from the aldehyde, and the adjacent quaternary center of **10** are formed with >99% stereocontrol.

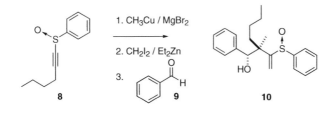

Preparation of Cyclic Amines

June 28, 2004

As cyclic amines are at the heart of medicinal chemistry, there is always interest in new methods for their preparation. Marco Ciufolini of the Université Claude Bernard in Lyon reports (Organic. Lett. 5:4943, 2003) the preparation of a series of dihydro indole derivatives, exemplified here by **3**, **6**, and **9**, by free radical cyclization of an N-O precursor. The N-O precursor can be prepared from the corresponding bromide, as illustrated by the conversion of **1** to **2** and of **4** to **5**. Alternatively, a radical precursor such as **8** can be prepared separately. The generated radical is then trapped by **7** to make a new radical, that cyclizes to **9**.

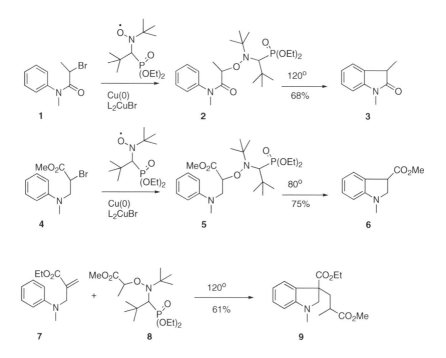

Heteroaromatics such as **10** are inexpensive compared to enantiomerically-pure cyclic amines such as **11**. Yong-Gui Zhou of the Dalian Institute of Chemical Physics reports (J. Am. Chem. Soc. 125: 10536, 2003) the development of a chiral Ir catalyst that effects hydrogenation of **10** to **11** (700 psi H$_2$, RT, 18 h) in 93% ee. The process is compatible with esters, alcohols and ethers on the sidechain.

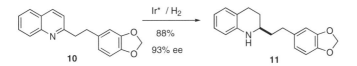

The Fischer indole synthesis has been a workhorse of medicinal chemistry. What has been needed is a procedure of comparable ease and efficiency for converting a ketone or aldehyde such as **12** to the corresponding pyridine, such as **13**. Antonio Arcadi of the University of Milan has now developed (J. Org. Chem. 68: 6959, 2003) just such a procedure, based on a gold-catalyzed condensation with propargylamine. The gold catalyst is commercially available. The regioselectivity of this procedure is noteworthy.

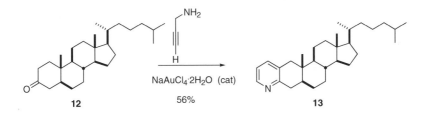

Enantioselective Total Synthesis of (+)-Amphidinolide T1

July 5, 2004

Amphdinolide T1 **1** is representative of a family of macrolides, isolated from the *Amphidinium* marine dinoflagellates, that show significant antitumor properties. Arun Ghosh of the University of Illinois at Chicago recently completed (J. Am. Chem. Soc. 125: 2374, 2003) a total synthesis of **1**, based conceptually on the convergent coupling of the enantiomerically-pure fragments **2** and **3**.

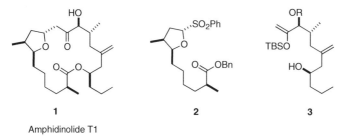

Amphidinolide T1

For each of the two fragments, a key component was assembled by the syn selective aldol condensation developed by Ghosh. For **2**, addition to 3-benzyloxypropionaldehyde gave **4**, which was carried on to the protected lactol **6**. Homologation to **7** allowed Grubbs coupling with the fragment **8**, leading to **9**. Activation of the lactol by condensation with benzenesulfinic acid

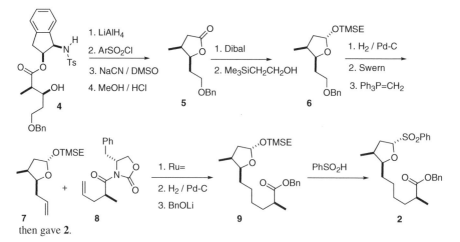

then gave **2**.

The enantiomerically-pure aldehyde **14** was prepared by adding dithiane to the commercially-available glycidyl tosylate **10**. For the other half of **3**, another syn-selective aldol condensation gave **12**, which was carried on to the iodide **13**. Reduction with *t*-butyl lithium, addition of the resulting organolithium to **11** and oxidation then gave the coupled ketone, which was homologated using the Petasis procedure to give **14**.

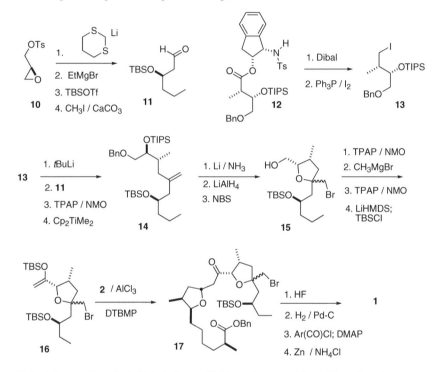

In fact, the sensitive disubstituted alkene of **14** turned out to not be stable to the subsequent AlCl₃ coupling conditions, so the alkene and the secondary alcohol were protected together as the bromoether **15**. Condensation of the derived enol ether **16** with the sulfone **2** in the presence of DTBMP (2,6-di-*t*-butyl-4-methylpyridine) then gave **17**. Yamaguchi lactonization followed by regeneration of the alkene by zinc reduction completed the synthesis of **1**.

Stereocontrolled Construction of Carbacycles

July 12, 2004

Intramolecular carbene insertion (e.g. **1** → **3**) has long been a useful method for ring construction. Masahisa Nakada of Waseda University in Tokyo now reports (J. Am. Chem. Soc. 125: 2860, 2003) that with the addition of the ligand **2** this process can be made highly enantioselective. As the starting diazo ketone **1** is easily prepared by diazo transfer to the sulfonyl ketone, this should allow facile entry to enantioenriched cyclopentanones and cyclohexanones.

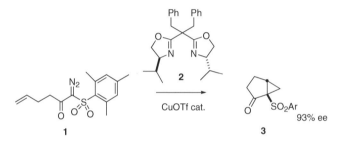

An even more common method for carbacyclic ring construction is the Diels-Alder reaction. Mukund Sibi of North Dakota State University reports (J. Am. Chem. Soc. 125: 9306, 2003) that the flexible ligand **6** works particularly well in mediating the enantioselective addition of **4** to **5**, to give **7**.

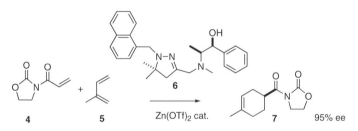

Another way to approach the enantioselective construction of carbocycles is to start with a readily-available carbohydrate. Gloria Rassu of the Insituto di Chimica Biomolecolare del CNR, Sassari, and Giovanni Casiraghi of the Università di Parma report (J. Org. Chem. 68: 5881, 2003) that the lactone **8** undergoes smooth aldol condensation to give the highly-substituted, and enantiomerically-pure, lactone **9**. The cyclization works equally well with the lactam in place of the lactone. Eight-membered rings can also be efficiently prepared using this approach.

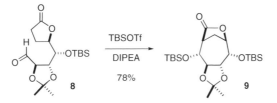

Intramolecular alkylation, although it is enticing, has not been developed as a method for cyclohexanone construction. Joseph P.A. Harrity of the University of Sheffield reports (J. Org. Chem. 68: 4392, 2003) that TiCl$_4$ smoothly transforms the enol ether **10**, prepared from the corresponding alkynyl phosphonium salt, into the 2-aryl cyclohexanone **11**. Alkynyl ethers such as **10** are readily prepared in enantiomerically-enriched form. Would the enantiomeric excess be maintained on cyclization?

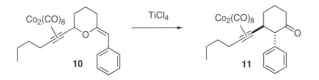

"Organometallic" Coupling without the Metal!

July 19, 2004

Ususally, an aryl halide such as **1** will be coupled with the arylboronic acid **2** using a Pd catalyst. Nicholas Leadbeater of King's College, London, reports (J. Org. Chem. 68: 5660, 2003) that the coupling can be carried out in water with microwave heating, with *no transition metal catalyst!* A wide range of aryl bromides and aryl boronic acids participate efficiently in this coupling. [Note: this was later retracted, J. Org. Chem. 70: 161, 2005.]

78%

Carbon-carbon bond formation between sp^3-hybridized carbons is not easy even with organometallic reagents. John Vederas of the University of Alberta reports (Organic Lett. 5: 2963, 2003) that UV irradiation at low temperature of diacyl peroxides such as **4** gives the coupled product **5**. The diacyl peroxides can be prepared by the DCC-mediated condensation of acids with peracids,

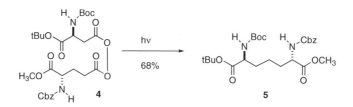

The alcohol **10** looks like it might be formed by the addition of a Grignard reagent to an aldehyde. In fact, Patrick Steel of the University of Durham prepared **10** (Tetrahedron Lett. 44: 9135, 2003) by Diels-Alder addition of the transient silene derived from **7** to the diene **8**. More highly substituted dienes lead to more complex arrays of stereogenic centers. The intermediate silacyclohexenes, exemplified by **9**, should also engage in the other reactions of allyl silanes.

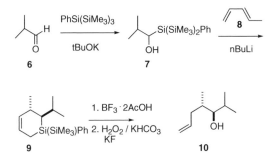

Preparation of Enantiomerically-Pure Building Blocks

July 16, 2004

Convergent construction directed toward the preparation of enantiomerically-pure targets depends on the availability of enantiomerically-pure starting materials. There are many ways that these can be prepared. One way is from carbohydrate precursors, but this approach has been limited, in that L-sugars are much more expensive than D-sugars. Tzenge-Lien Shih of Tamkang University in Taipei describes (Tetrahedron Lett. 45: 1789, 2004) the facile conversion of the inexpensive D-ribonolactone derivative **1** to the much more valuable L-ribonolactone derivative **3**, by careful hydrolysis of the intermediate mesylate. The epoxide **2** is presumed to be an intermediate in this transformation.

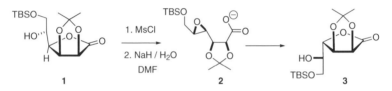

While secondary alcohols are now relatively easy to prepare in enantiomerically-pure form, secondary amines have been more challenging. Larry Overman of UC Irvine reports (J. Am. Chem. Soc. 125: 12412, 2003) the catalytic rearrangement of primary allylic alcohols such as **4** to the corresponding protected vinyl amine **5** with excellent ee. Hydrolysis of the amine **5** gives the GABA aminotransaminase inhibitor **6**. Unnatural amino acids can be prepared by oxidative cleavage of the protected vinyl amines.

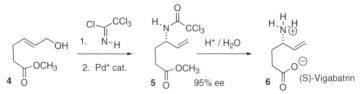

Another approach to secondary amines has been reported (J. Am. Chem. Soc. 125: 16178, 2003) by Masakatsu Shibasaki of the University of Tokyo. Addition of methoxyamine to a chalcone **7** (alkyl enones work in slightly lower ee) gives the amine **8**. The amine **8** can be reduced with high stereocontrol to the amino alcohol **9**. K-Selectride gives the complementary diastereomer.

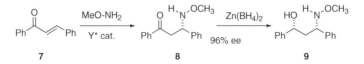

56

Professor Shibasaki has also investigated (J. Am. Chem. Soc. 125: 15840, 2003) Michael addition to prepare alkylated secondary centers in high enantiomeric excess. Addition of substituted acetoacetates to cyclohexenone and to cycloheptenone proceeds with high ee. With the more reactive cyclopentenone, the ee is slightly lower.

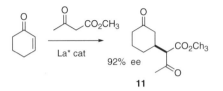

11

It is apparent that "enantiomerically-pure", written over and over again, can be cumbersome. We have suggested (C&E News Aug. 19, 1991, p. 5; J. Org. Chem. 57: 5990, 1992) "symchiral" as a pleasing alternative.

Synthesis of (-)-Strychnine

August 2, 2004

The total synthesis of (-)-strychnine **3** reported (J. Am. Chem. Soc. 125: 9801, 2003) by Miwako Mori of Hokkaido University is a *tour de force* of selective organopalladium couplings.

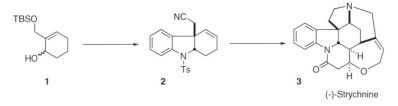

1   2   3

(-)-Strychnine

The absolute configuration of the final product was established at the outset, by Pd-catalyzed de-racemizing coupling of **4** with the *o*-bromoaniline derivative **5**. Using the inexpensive Binol-derived ligand (S)-BINAPO **6** , a model coupling was carried out on several cyclohexenol derivatives having different one-carbon substituents at C-2. The best ee's were observed with the silyloxymethyl group. Several alcohol derivatives were then tried, and it was found that the allylic phosphate gave the best rates and ee's. Using the optimized **4**, the coupling with **5** to give **7** proceeded in 84% ee.

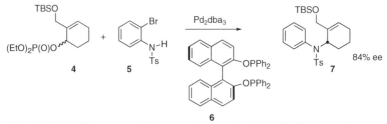

4   5   7   84% ee

6

The sidechain of **7** was extended by one carbon, to give the nitrile **8**. A second organopalladium step then was used to cyclize **8** to **2**. Using Ag$_2$CO$_3$ as the base suppressed unwanted alkene migration. The reaction ran more slowly in DMSO than in DMF, but byproduct formation was suppressed.

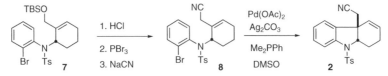

7   1. HCl   8   Pd(OAc)$_2$ 2

2. PBr$_3$ Ag$_2$CO$_3$

3. NaCN   Me$_2$PPh

DMSO

The crystalline nitrile **2** (99% ee from EtOH) was reduced and protected to give the carbamate **9**, setting the stage for another Pd-catalyzed ring-forming step. Allylic oxidation of **9** gave the enone only in unacceptably low yield. Pd-mediated cyclization, by contrast, proceeded efficiently to give the alkene **10**. Hydroboration followed by oxidation then gave the ketone **11**, a useful intermediate for the construction of a variety of *Strychnos* alkaloids.

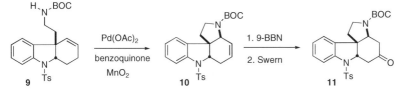

For strychnine **3**, the ketone **11** was converted to the alkene **12** by reduction of the enol triflate derived from the more stable enolate. Deprotection and acylation gave **13**, which was cyclized with Pd to give, after equilibration, the diene **14**. Alkylation, to give **15**, followed by Pd-mediated cyclization then gave **16**, which was reduced and cyclized to (-)-strychnine **3**.

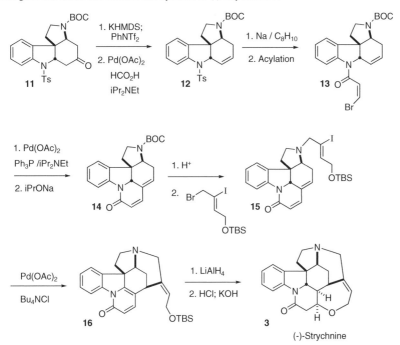

Pd-Mediated Coupling in Organic Synthesis: Recent Milestones

August 9, 2004

Many aryl coupling reactions have been carried out on bromides, but often the much more expensive aryl triflates are required. Pierre Vogel of the Swiss Institute of Technology in Lausanne has carried out (J. Am. Chem. Soc. 125: 15292, 2003) a detailed investigation of the Stille coupling with a series of inexpensive arenesulfonyl chorides, including **1**. In addition to carbonylative coupling, to give **4**, and non-carbonylative coupling, to give **5**, the reaction can be directed toward the thioester **6**. This is a new and potentially very useful procedure for reducing an arenesulfonate to the protected thiophenol.

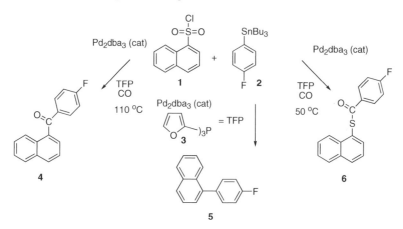

Alkynes are usually alkylated under strongly basic conditions. Gregory Fu of MIT recently reported (J. Am. Chem. Soc. 125: 13642, 2003) a much milder Pd and Cu mediated coupling, illustrated by the reaction of the terminal alkyne **7** with the alkyl iodide **8** to give **9**. Alkyl bromides work equally well. It is exciting that ketones, esters, alkyl chlorides, alkenes, nitriles and acetals are compatible with the procedure. The key to the reaction is the use of the supporting carbene ligand **10**.

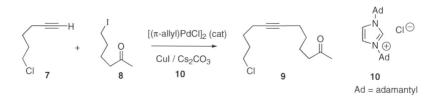

Ad = adamantyl

Professor Fu has also developed (J. Am. Chem. Soc. 125: 13642, 2003) an equally powerful Pd-mediated procedure for sp^3- sp^3 coupling. With **14** as the supporting phosphine, the organozinc bromide **11** (easily prepared by the action of Zn metal on the bromide) couples with the bromide **12** to give **13**. Chlorides and tosylates also serve efficiently as leaving groups, and ethers, amides and acetals are compatible with the coupling conditions. The organozinc halide and/or the coupling partner may also be sp^2 hybridized. The coupling reaction is limited to the formation of *primary* sp^3-hybridized bonds.

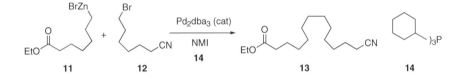

| **11** | **12** | **13** | **14** |

Enantioselective C-C Bond Construction:
Part One of Three

August 16, 2004

The enantioselective addition of allyl organometallics to carbonyls has become one of the workhorses of organic synthesis. Dennis Hall of the University of Alberta reports (J. Am. Chem. Soc. 125: 10160, 2003) the scandium triflate catalysis chiral allylboronic acids become more effective tools. The best of these, the Hoffmann camphor derivative **2**, adds to aldehydes under $Sc(OTf)_3$ catalysis with excellent enantiomeric excess. The reaction works equally well for methallyl, and for the *E* and *Z* crotyl boronic acids. The crotyl derivatives react with the expected high diastereocontrol. A limitation to the boronate additions is that branched chain aldehydes give low yields.

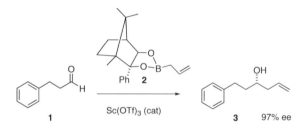

A complementary method for allylation reported (J. Org. Chem. 68: 5593, 2003) by Hishashi Yamamoto, now of the University of Chicago, works *particularly well with branched aldehydes*. The allyl source is the inexpensive allyltrimethoxysilane, with BINAP-complexed Ag ion as the catalyst. Activation of the allylsilane with KF and 18-crown-6 is critical to the success of this reaction.

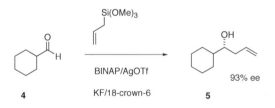

Thomas Lectka of Johns Hopkins University has reported (J. Org. Chem. 68: 5819, 2003) that benzoylquinine (BQ) catalyzes the two-carbon homologation of a ketene, derived from the acid chloride, with chloroamide such **7**, to give the β-amino acid derivative **8** with control of both relative and absolute configuration. The authors suggest that the BQ is involved five times in the course of the transformation of **6** into **8**. The two esters of the product are differentiated, so one can imagine, inter alia, reduction of one or the other to the alcohol, and formation of the activated aziridine or azetidine. These could then be further homologated.

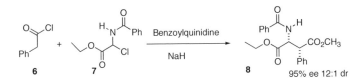

As reported (J. Org. Chem. 68: 6197, 2003) by Yoshjii Takemoto of Kyoto University, α-amino acids can be prepared in high enantiomeric and diastereomeric excess by Ir-mediated two-carbon homologation of allylic phosphates such as **9** with the protected glycine **10**. Either diastereomer can be made dominant by varying the reaction conditions.

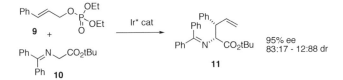

Enantioselective C-C Bond Construction: Part Two of Three

August 23, 2004

Stilbene diols such as **3** are gaining prominence both as synthetic intermediates and as effective chiral auxiliaries. While the diols can be prepared in high ee by Sharpless dihydroxylation, it would be even more practical to prepare them by direct asymmetric pinacol coupling. N. N. Joshi of the National Chemical Laboratory in Pune reports (J. Org. Chem. 68: 5668, 2003) that 10 mol % of the inexpensive Ti salen complex **2** is sufficient to effect highly enantioselective and diastereoselective pinacol coupling of a variety of aromatic aldehydes. Most of the product diols are brought to >99% ee by a single recrystallization.

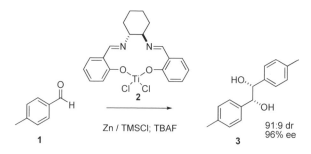

The coupling of the racemic allylic acetate **4** with malonate can give either the terminal product **5** or the internal product **6**. Tamio Hayashi of Kyoto University reports (Organic Lett. 5: 1713, 2003) that using a Rh catalyst in the presence of Cs_2CO_3 and a chiral phosphine leads to a 1:99 ratio in favor of the internal product **6**, with outstanding ee.

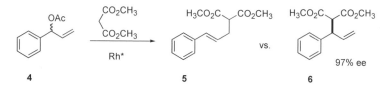

Starting with the racemic carbonate **7** and using a Mo catalyst, Christina Moberg of the Royal Institute of Technology (KTH) in Stockholm was able to achieve (Organic Lett. 5: 2275, 2003) 26:1 regioselectivity in favor of the branched product **9**, again with outstanding ee. In this case, the pyridylamide ligand for the Mo is polymer-bound, so it is easily recycled. Remarkably, this high ee was observed for reactions run at elevated temperature with microwave promotion (6 minutes, 160°).

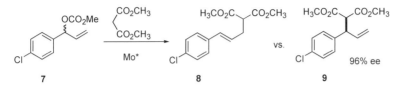

7   8   vs.   9   96% ee

Albert S.C. Chan of the Hong Kong Polytechnic University reports (J. Org. Chem. 68: 1589, 2003) two important transformations. The three-component (Mannich) condensation of **10** with **11** and **12** proceeds with high diastereoselectivity, to give the amino alcohol **13**. Hydroboration of the alkyne **14** followed by transmetalation of the intermediate vinyl borane gives a zinc species, which under catalysis by the easily-prepared β-naphthol **13** adds to aromatic and branched aldehydes with high ee. The product allylic alcohols are useful intermediates for organic synthesis.

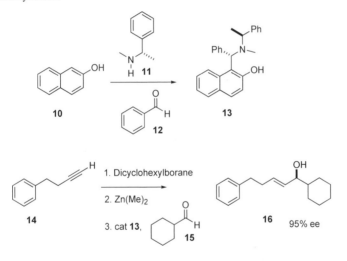

Enantioselective C-C Bond Construction:
Part Three of Three

August 30, 2004

Sulfones are chemical chameleons, electron-withdrawing groups that are also good leaving groups. Tamio Hayashi of Kyoto University took advantage of this in designing (J. Am. Chem. Soc. 125: 2872, 2003) an enantioselective method for the construction of ternary stereogenic centers. The Rh-catalyzed conjugate addition of the aryl Ti species **2** to the unsaturated sulfone proceeds with high enantioselectivity. Subsequent β-hydride elimination proceeds, as expected, away from the newly-formed ternary center. Readdition of Rh-H followed by reductive elimination of the sulfone then gives the alkene **3**.

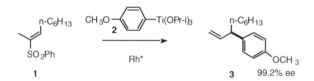

There has been a continuing effort to make the Baylis-Hillman reaction a catalytic asymmetric process. Scott Schnauss of Boston University recently reported (J. Am. Chem. Soc. 125: 12094, 2003) an elegant solution to this problem, based on the use of Binol-derived Bronsted acids as catalysts. The product hydroxy enones such as **6** are interesting in themselves, and also as substrates for further transformation, for instance by Claisen rearrangement.

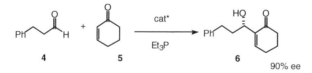

Prochiral α-substituted enones such as **7** are inexpensive starting materials. Patrick Walsh of the University of Pennsylvania recently reported (J. Am. Chem. Soc. 125: 9544, 2003) a catalytic enantioselective procedure for the 1,2-addition of dialkyl zinc reagents to such enones. The chiral catalyst is a sulfonamide derived from 1,2-diaminocyclohexane. The tertiary allylic alcohols are useful products, difficult to prepare by other methods. Even more exciting is the observation that addition of oxygen to the reaction mixture directly converts the tertiary alkoxide to the epoxide **9** with high diastereocontrol. Subsequent Lewis acid-catalyzed rearrangement of the epoxide **9** then gives the ketone **10**. The overall process sets the absolute configuration of two stereogenic centers. The construction of cyclic quaternary stereogenic centers is particularly noteworthy.

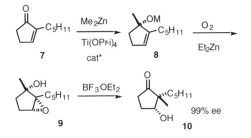

Synthesis of (-)-Podophyllotoxin

September 6, 2004

(-)-Podophyllotoxin **1** and its derivative etoposide **2**, derived from natural sources, are in current clinical use. Michael Sherburn of Australian National University reports (J. Am. Chem. Soc. 125:

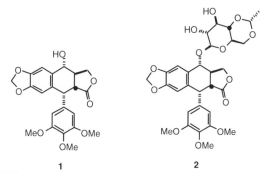

1 **2**

12108, 2003) complementary total syntheses of **1** and of its enantiomer,. The key step in each of these syntheses is a spectacular intramolecular alkene arylation, exemplified by the conversion of **4** to **5**.

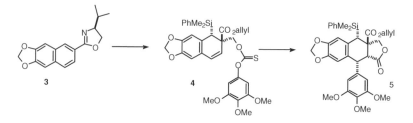

3 **4** **5**

The absolute configuration of **1** was set by conjugate addition to the oxazoline **3** followed by trapping of the product anion, following the precedent of Meyers. Reduction of the oxazoline to the alcohol followed by thionocarbonate formation then set the stage for the key aryl transfer reaction.

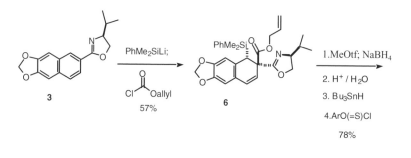

3 **6**

PhMe₂SiLi;

Cl—Oallyl

57%

1. MeOtf; NaBH₄
2. H⁺ / H₂O
3. Bu₃SnH
4. ArO(=S)Cl

78%

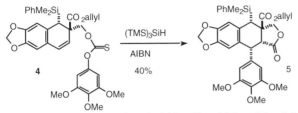

4   (TMS)₃SiH   AIBN   40%   **5**

The aryl transfer is thought to be initiated by addition of the silyl radical to the thiocarbonyl. The radical so formed adds to the alkene to generate the benzylic radical **7**. This radical adds to the arene to give **8**, which fragments to **5**.

Oxidation of the silyl group of **5** to the ketone followed by Pd-mediated decarboxylation led to the ketone **9**. The conversion of **9** to podophyllotoxin **1** followed the literature precedent.

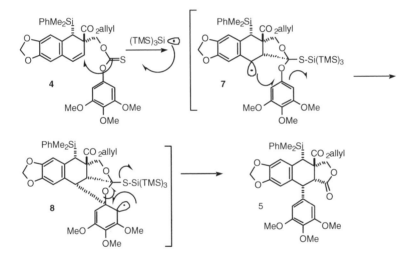

The Grubbs Reaction in Organic Synthesis: Part One of Three

September 13, 2004

We last reviewed organic synthesis applications of the Grubbs reaction on April 19, 2004. The (relatively) robust nature of the commercially-available catalyst and its commercial availability have spurred the expanding exploration of the scope of this reaction. Through this month, we will feature some recent highlights.

The most straightforward application of the Grubbs reaction is to effect homologation of a terminal vinyl group. One concern with the use of the second generation Grubbs catalyst **3** is the cost, about $100/mmol. In conjunction with a total synthesis of the macrolide RK-397, Frank McDonald of Emory reported (J. Am. Chem. Soc. 126: 2495, 2004) that the conversion of **1** to **2** required 10 mol % of **3**, but that it proceeded efficiently with just 2 mol % of the Hoveyda Ru catalyst **4** (J. Am. Chem. Soc. 122: 8168, 2002). The alkene so prepared was cleanly trans. An advantage of **4** is that it avoids the use of the expensive PCy$_3$.

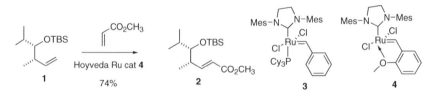

Because of the functional group tolerance of the Grubbs catalyst **3**, it will operate even with complex substrates. In the course of a total synthesis of (-)-cytisine, Giordano Lesma and Alessandra Silvani of the University of Milan reported (Organic Lett. 6: 493, 2004) that **5**, with a free N-H, could be cyclized efficiently to **6**.

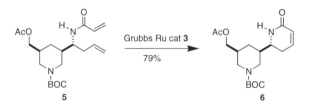

The even more spectacular cyclization of **7** to Arenastatin A **8** was reported (Tetrahedron Lett. 45: 5309, 2004) by Gunda Georg of the University of Kansas. In this case, the catalyst used was the first generation Grubbs catalyst, **9**.

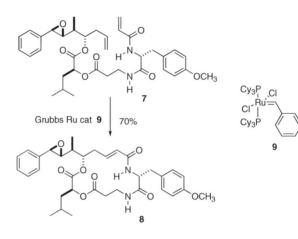

Grubbs Ru cat **9**   70%

The Grubbs Reaction in Organic Synthesis: Part Two of Three

September 20, 2004

We last reviewed organic synthesis applications of the Grubbs reaction on April 19, 2004. The (relatively) robust nature of the commercially-available catalyst and its commercial availability have spurred the expanding exploration of the scope of this reaction. This month, we are featuring some recent highlights.

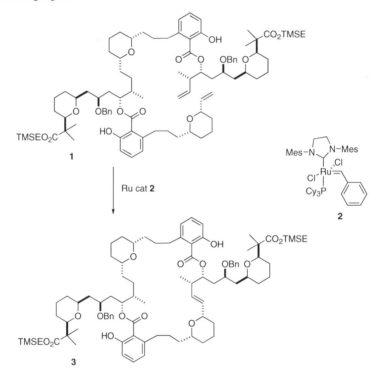

There has been a great deal of effort over the past several years directed toward the total synthesis of physiologically-active macrolactones and macrolactams. The availability and functional group tolerance of the Grubbs catalysts, especially the second generation Grubbs catalyst **2**, now make alkene metathesis an attractive procedure for closing the macrocyclic ring. Eun Lee of Seoul National University, in the course of a total synthesis of (+)-SCH 351444, a novel activator of LDL-R promoter, reported (*J. Am. Chem. Soc.* **2004**, *126*, 2680) the cyclization of **1** to **3**.

Other alkenes in the substrate, and especially other alkynes, will react with the Ru catalyst. Samuel Danishefsky, of Columbia University and Sloan-Kettering Institute, recently completed (*Organic Lett.* **2004**, *6*, 413) the total synthesis of the antimalarial and antitumor agent aigialomycin D. For the Ru-mediated macrocyclization of **4** to **5**, the alkyne in the molecule was protected as the dicobalt octacarbonyl adduct. After the cyclization, the alkyne was deprotected by brief exposure to ceric ammonium nitrate.

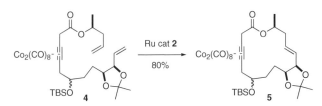

Selectivity can sometimes be achieved by changing the Ru catalyst. James Panek of Boston University, reported (*Organic Lett.* **2004**, *6*, 525) that the attempted conversion of **6** to the tetraene macrolactam core **7** of the cyclotrienins led instead to the unwanted **8**. Use of the less-reactive first generation Grubbs catalyst **9** gave clean cyclization to the desired **7**.

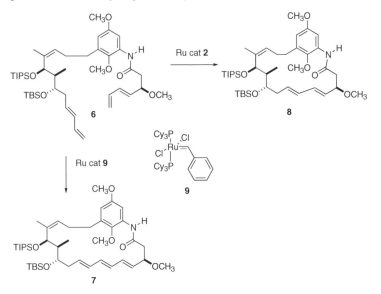

The Grubbs Reaction in Organic Synthesis: Part Three of Three

September 27, 2004

We last reviewed organic synthesis applications of the Grubbs reaction on April 19, 2004. The (relatively) robust nature of the commercially-available catalyst and its commercial availability have spurred the expanding exploration of the scope of this reaction. This month, we are featuring some recent highlights.

A simple yet powerful application of the Grubbs reaction is specific homologation of a terminal vinyl group. When there is more than one alkene in the molecule, suitably disposed, one would worry about competing cyclization. In studies directed toward the cryptophycins, Mark Lautens of the University of Toronto has reported (*Organic Lett.* **2004**, *6*, 1883) that the second generation Grubbs catalyst **2** smoothly converts **1** to **3**. The alternative cyclization product **4** is produced only in trace amounts.

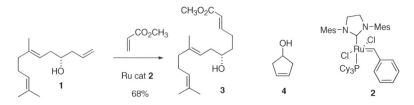

As illustrated above, free alcohols are compatible with the Grubbs reaction. In the course of a synthesis of (+)- puraquinonic acid, Derrick Clive of the University of Alberta reported (*J. Org. Chem.* **2004**, *69*, 4116) that the very easily oxidized alcohol **5** maintains its enantiomeric excess as it is cyclized with the catalyst **2** to give **6**.

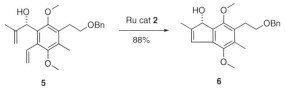

The Grubbs reaction is an equilibration, and so would not be expected to be generally effective for the preparation of medium rings. Radomir Saicic of the University of Belgrade, Serbia and Montenegro, has reported (*Organic Lett.* **2004**, *6*, 1221) an elegant solution to this problem. The diene **7** is easily prepared by alkylation of cyclopentanone (or cyclohexanone) carboxylate, followed by the addition of allyl magnesium chloride. Exposure to the first-generation Grubbs catalyst **8** easily forms the six-membered ring, to give **9**. Subsequent Grob fragmentation then delivers **10**. Nine, ten, and eleven-membered rings were prepared using this approach.

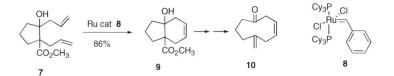

Enyne metathesis can also be used with highly substituted substrates. Catherine Lièvre of the Université de Picardie reports (*J. Org. Chem.* **2004**, *69*, 3400) that enynes such as **11**, readily prepared from carbohydrate precursors, are cyclized by the second generation Grubbs catalyst **2** to the enantiomerically-pure cyclic dienes, exemplified by **12**.

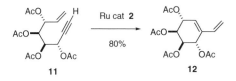

Enyne metathesis can also be used to prepare more complex structures. A key step in the synthesis of (+)-viroallosecurinine reported (*Tetrahedron Lett.* **2004**, *45*, 5211) by Toshio Honda of Hoshi University in Tokyo is the selective cyclization of **13** to **15**. In this case, the nitro Hoveyda-type catalyst **14** was used.

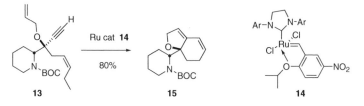

Applications of the Grubbs reaction continue to be developed. This is an update, not the final word.

Synthesis of Deacetoxyalcyonin Acetate

October 4, 2004

Deacetoxyalcyonin acetate **1** and euncellin **2** are representative members of the eunicellin class of diterpenes. The synthesis of deacetoxyalcyonin acetate **2** by Gary Molander of the University of Pennsylvania (*J. Am. Chem. Soc.* **2004**, *126*, 1642) illustrates the power of intramolecular organometallic carbonyl addition for ring construction.

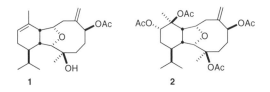

The six-membered ring of **1** was commerically available in enantiomerically-pure form as α-phellandrene **3**. The challenge was to stitch the highly-substituted ten-membered ring of **1** onto the disubstituted alkene of **3**. The strategy that was conceived was to first construct the seven-membered ring of **8**, then effect three-carbon ring expansion to give **11**.

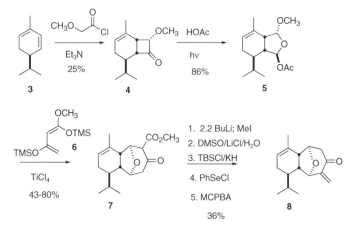

The plan for seven-membered ring construction was to effect stepwise 4 + 3 cycloaddition of **6** to the protected dialdehyde **5**. The preparation of **5** began began with 2+2 cycloaddition between **3** and methoxy ketene, to give **4** with high regio- and diastereocontrol. Photochemical cleavage then gave **5**. The acid-mediated 4 + 3 proceeded via initial addition to the less congested ionized aldehyde, to give the β-keto ester **7**. Alkylation of the dianion of **7** followed by ester hydrolysis and selenation /oxidation established the enone **8**.

Three-carbon ring expansion was carried out in two stages. First, two-carbon homologation of the exo methylene ketone **8** followed by trapping of the intermediate enolate as the triflate led to **9**. Nozaki-Hiyama-Kishi coupling followed by acetylation smoothly converted **9** into **10**.

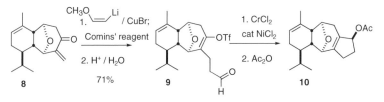

The trisubstituted alkene of **10** was more readily oxidized than was the congested tetrasubstituted alkene, so the more reactive alkene was temporarily epoxidized. After ozonolysis, the epoxide was reduced off using the Sharpless protocol. It is a tribute to the specificity of this reagent that the easily-reduced α-acetoxy ketone is not affected. Selective silylation of the more accessible ketone followed by methylenation, hydrolysis and addition of methyl lithium to the outside face of the previously protected carbonyl then delivered **1**.

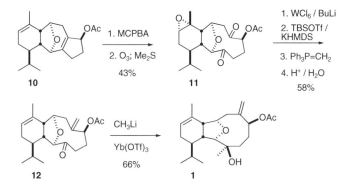

Enantioselective Ring Construction: Part One of Three

October 11, 2004

As new drug entities must be usually be prepared as single enantiomers, and as many contain one or more heterocyclic or carbocyclic rings, there is an increased emphasis on the development of practical methods for the construction of enantiomerically pure cyclic systems. In this three-part series, we will cover the most important recent advances.

One of the most powerful strategies for asymmetric ring construction is to desymmetrize a preformed ring. Yasamusa Hamada of Chiba University in Japan has reported (*J. Am. Chem. Soc.* **2004**, *126*, 3690) that the inexpensive diaminophosphine oxide **2** nicely catalyzes the asymmetric alkylation of the cyclohexanone carboxylate **1** to give **3**. Although no examples were given, this asymmetric alkylation would probably work as well with heterocyclic β-ketoesters.

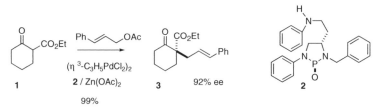

While enantioselective transition metal catalysis continues to be important, several useful all-organic catalysts have been developed over the past few years. Tomislav Rovis of Colorado State University has reported (*J. Am. Chem. Soc.* **2004**, *126*, 8876) that the triazolium salt **5** catalyzes the enantioselective Stetter-type cyclization of **4** to **6**. The cyclization also works well for the enantioselective construction of azacyclic, thiacyclic and carbocyclic rings.

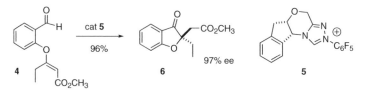

Another example of organic catalysis was reported (*J. Am. Chem. Soc.* **2004**, *126*, 450) by Benjamin List of the Max Planck Institute, Mülheim. The amino acid **8** cyclizes **7** to **9** efficiently and with high enantioselectivity. This is particularly remarkable given the ease with which **9** would be expected to racemize under acidic or alkaline conditions.

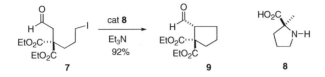

Enantioselective Ring Construction: Part Two of Three

October 18, 2004

As new drug entities must be usually be prepared as single enantiomers, and as many contain one or more heterocyclic or carbocyclic rings, there is an increased emphasis on the development of practical methods for the construction of enantiomerically pure cyclic systems. In this three-part series, we will cover the most important recent advances.

Enantiomerically-pure ternary stereogenic centers, often readily available, can be used to set the helicity of ring-forming reactions, leading to diastereomerically- and so enantiomerically-pure products. In the course of a synthesis of the antifungal sesquiterpene (-)-alliacol A **5**, Kevin Moeller of Washington University (*J. Am. Chem. Soc.* **2004**, *126*, 9106) set the absolute configuration of **2** by chiral catalytic conjugate addition to the prochiral **1**. Electrochemical oxidation of **2** gave **3**, which was cyclized to **4**. Further manipulation of **4** gave **5**.

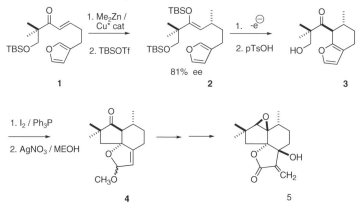

Another approach to the preparation of enantiomerically-pure ring systems is to start with the prochiral preformed ring, and perform an asymmetric transformation. Mauro Pineschi of the Università di Pisa has reported (*J. Org. Chem.* **2004**, *69*, 2099) a study of a series of racemic cyclic epoxides, represented here by **6**. Exposure of **6** to Bu₂Zn in the presence of an enantiomerically-pure Cu complex derived from **7** converts one enantiomer of **6** to **8**, and the other enantiomer to **9**, each in high enantiomeric excess. Several other classes of cyclic epoxides showed similar selectivity.

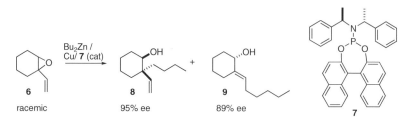

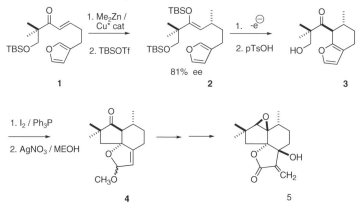

80

Epoxides such as **10** can be prepared in high enantiomeric purity, by, *inter alia*, kinetic resolution. David Hodgson of Oxford University has demonstrated (*J. Am. Chem. Soc.* **2004**, *126*, 8664) that on exposure to LTMP, monosubstituted epoxides are smoothly converted into the corresponding alkoxy carbene or alkoxy carbenoid. When an alkene is available for insertion, the cyclopropane, in this case **11**, is formed with high diastereocontrol. This represents a powerful new approach to enantioselective ring construction. It is possible that in the absence of a target alkene, the intermediate alkoxy carbene could divert to intramolecular C-H insertion, which might also proceed with substantial diastereocontrol.

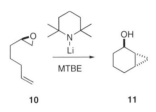

       10   **11**

Alkyne Metathesis in Synthesis: Syntheses of (+)-Ferrugine and Anatoxin-α

October 25, 2004

Although much emphasis has been placed on the importance of alkene metathesis in organic synthesis, alkyne metathesis is also significant. Metathesis reactions of alkynes with alkenes and metathesis reactions of alkynes with alkynes have both been carried out efficiently.

An enantiospecific synthesis of the potent anti-Alzheimer's agent (+)-ferruginine **1** and two enantiospecific syntheses of the potent anti-Alzheimer's agent anatoxin-α **2** were submitted essentially simultaneously earlier this year. In each case, the starting material used was the inexpensive pyroglutamic acid **3**, and in each case the synthesis depended on Ru-catalyzed alkyne-alkene metathesis.

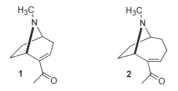

The first report, received on February 25th, by Varinder Aggarwal of Bristol University (*Organic Lett.* **2004**, *6*, 1469) is representative of all three. Ohira homologation of the aldehyde **6** gave the enyne

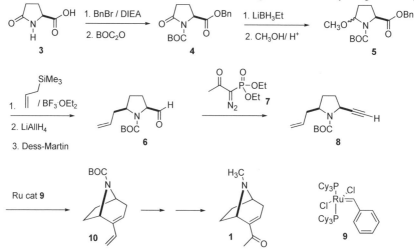

8. A detailed study showed that the alkyne-alkene metathesis was proceeding cleanly, but that the product **10** was then decomposed by the Ru catalyst. Use of the less reactive first generation Grubbs catalyst **9** gave clean conversion of **8** to **10**.

The second report (*Tetrahedron Lett.* **2004**, *45*, 4397), received on February 27th, was from Miwako Mori at Hokkaido University. The cyclization of **11** to form the seven-membered ring gave only low yields. Cyclization of the silyl derivative **12**, in contrast, proceeded efficiently, with concomitant desilylation.

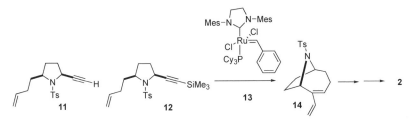

The third report (*Organic Lett.* **2004**, *6*, 1329), received on February 29th, was from Stephen Martin at the University of Texas. The starting material in this case was the other enantiomer of pyroglutamic acid **3**, to deliver the natural enantiomer, (+)-anatoxin-α **2**. Metathesis was carried out on the eneyne **15**, again using the second-generation Grubbs catalyst **13**. Selective oxidative cleavage then led to **2**.

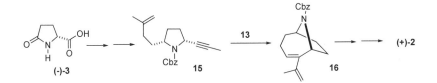

Mark Overhand of Leiden University recently reported (*Tetrahedron Lett.* **2004**, *45*, 4379) an example of alkyne-alkyne metathesis, the cyclization of **17** to **19**. For this reaction, a tungsten catalyst was used.

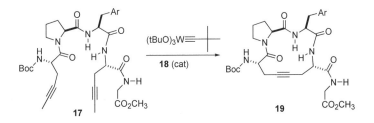

Catalytic Asymmetric Synthesis of Quinine and Quinidine

November 1, 2004

The tetracyclic alkaloid quinine **1** and the diastereomeric alkaloid quinidine **2** share a storied history. Eric Jacobsen of Harvard recently completed (*J. Am. Chem. Soc.* **2004**, *126*, 706) syntheses of enantiomerically-pure **1** and of **2**. For each synthesis, the key reaction for establishing the asymmetry of the target molecule was the enantioselective conjugate addition developed by the Jacobsen group.

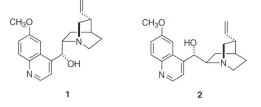

For both **1** and **2**, the synthesis started with the alkenyl amide **3**. Salen-mediated conjugate addition proceeded with remarkable induction, to give **5** in 92% ee as a mixture of diastereomers. Reduction and cyclization followed by deprotonation and kinetic quench delivered the enantiomerically-enriched cis dialkyl piperidine **6**. Homologation of the two sidechains then gave the alkenyl boronic ester **8**.

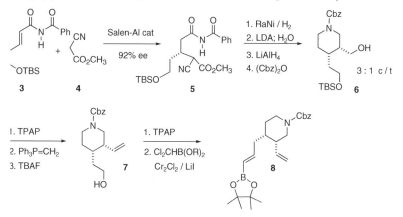

The quinoline portion of the target alkaloids was prepared by condensing p-anisidine **9** with ethyl propiolate, followed by bromination. Coupling of **10** with the boronic ester **8** proceeded to give **11**, the intermediate for the synthesis of both **1** and **2**. Selective direct epoxidation of **11** using the usual reagents failed, but Sharpless asymmetric dihydroxylation was successful, providing the diol in > 96:4

diastereoselectivity, with only traces of the tetraol and of the product from dihydroxylation of the terminal vinyl group. The diol could be converted cleanly to the desired epoxide **11**. Deprotection followed by cyclization led to quinine **1**. Preparation of the diastereomeric epoxide, using AD-mix-α, followed by cyclization gave quinidine **2**. The brevity of the preparation of these two classical alkaloids is a testament to the power of reagent-controlled synthesis.

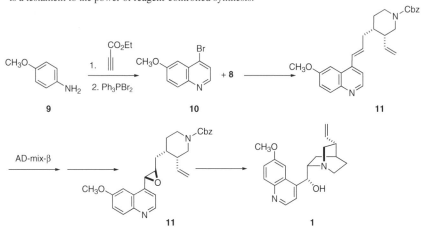

Best Synthetic Methods: Oxidation and Reduction

November 8, 2004

Although oxidation and reduction are common organic synthesis transformations, there is always room for improvement. The ideal reaction might use a solid, easily-.measured reagent, a minimal amount of solvent, need no work-up other than simple filtration, and proceed in > 95% yield. The zirconium borohydride piperazine complex reported (*Tetrahedron Lett.* **2004**, *45*, 3295) by M. Talbakksh and M.M. Lakouraj of Mazandaran University, Iran comes close to this ideal. The reagent is prepared by combining $ZrCl_4$ and an ethereal solution of $LiBH_4$ with piperazine at ice temperature. The resulting white powder is stable for months if stored dry ("vacuum dessicator"). The reagent reduces aldehydes in ether solution at room temperature. Ketones reduce in a few hours in refluxing ether, conditions that also reduce esters. Cyclohexenone **1** reduces cleanly to the allylic alcohol **2**, without the conjugate reduction often seen with other reducing agents. In each case, the reaction mixture was filtered and the solvent evaporated to give the pure product(s).

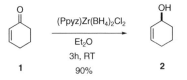

Direct reduction of an aldehyde or ketone to the corresponding ether could potentially telescope two reactions, reduction and protection, into one step. S. Chandrasekhar of the Indian Institute of Chemical Technology, Hyderabad, reports (*Tetrahedron Lett.* **2004**, *45*, 5497) that in the present of polymethylhydrosiloxane (PMHS) and catalytic $B(C_6F_5)_3$, TMS ethers of alcohols will convert aldehydes to the corresponding dialkyl ethers. The reaction works well for both saturated and benzylic alcohols. This may prove to be a useful alternative to Williamson synthesis for the preparation of complex ethers.

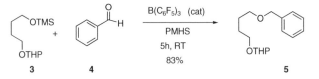

Hydrogen peroxide is an inexpensive oxidant, but it requires a catalyst to effect oxidation of an alcohol to the ketone. Removal of the catalyst then becomes an issue. Ronny Neumann of the Weizmann Institute of Science reports (*J. Am. Chem. Soc.* **2004**, *126*, 884) the development of a hybrid organic-tungsten polyoxometalate complex that is not soluble in organic solvents, but that nonetheless catalyzes the hydrogen peroxide oxidation of alcohols to ketones. The solid catalyst is removed by filtration after the completion of the reaction. The catalyst retained its activity after five recyles.

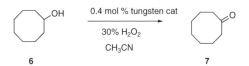

In the latest development of his elegant work with hydrazone derivatives, Andrew Myers of Harvard reports (*J. Am. Chem. Soc.* **2004**, *126*, 5436) that $Sc(OTf)_3$ catalyzes the addition of 1,2-bis(*t*-butyldimethylsilyl)hydrazine, to aldehydes and ketones to form the *t*-butyldimethylsilylhydrazones. Addition of tBuOH/tBuOK in DMSO to the crude hydrazone effects low temperature Wolff-Kishner reduction. Alternatively, halogenation of ketone hydrazones can lead to vinyl halides such as **11**, or the 1,1-dihalo derivatives, depending on conditions. Halogenation of aldehyde hydrazones provides the 1,1-dihalo derivatives such as **13**.

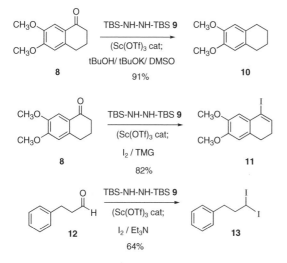

Best Synthetic Methods: Enantioselective Oxidation and Reduction

November 15, 2004

Usually, only a single enantiomer of a pharmaceutical is the useful drug. Often, the synthesis of the single enantiomer depends on the ability to form C-O or C-N bonds with high asymmetric induction. In the ideal asymmetric synthesis, *both* enantiomers of the starting material would be converted to the same enantiomer of the product. Mahn-Joo Kim and Jaiwook Park of Pohang University of Science and Technology, Korea report (*J. Org. Chem.* **2004**, *69*, 1972) an example of a simple solution to this problem. Selective acylation by a lipase is limited, because the increasing relative concentation of the less reactive residual enantiomer makes the esterification progressively less enantioselective. If the enantiomers of the starting alcohol could be interconverted, there would be no relative build-up of the less reactive enantiomer, and the enantioselectivity would be maintained. The authors report the development of an efficient Ru catalyst for the interconversion of the two enantiomers of **1** by in situ oxidation and reduction.

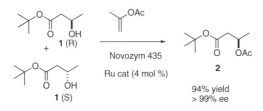

When the substrate is readily epimerizable, as with **3**, the problem is even easier. Xumu Zhang of Pennsylvania State University reports (*J. Am. Chem. Soc.* **2004**, *126*, 1626) the development of a chiral Ru complex (derived from C₃-Tunephos) that selectively hydrogenates one of the two interconverting enantiomers of **3**, delivering **4** with high enantioselectivity and diastereoselectivity.

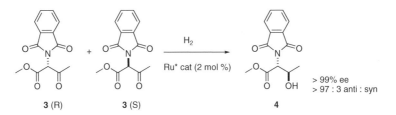

The Sharpless asymmetric dihydroxylation has played a prominent role in enantioselecitve organic synthesis. Two groups have recently reported improvements in the procedure. Osmo E.O. Horni of the University of Oulu, Finland has found (*J. Org. Chem.* **2004**, *69*, 4816) that sodium chlorite is a more efficient reoxidant than is the usual K₃[Fe(CN)₆]. Carlos A.M. Alfonso of the Instituto Superior Técnico, Lisbon has reported (*J. Org. Chem.* **2004**, *69*, 4381) that the asymmetric dihydroxylation can be carried out using an ionic liquid as the solvent. The product can be extracted with ether, leaving the osmium catalyst in the ionic liquid to be used again. Note that even monosubstituted alkenes such as **5** are dihydroxylated with high enantioselectivity with these systems.

Alcohol oxidation can also be enantioselective. The best systems reported to date for the selective oxidation of one enantiomer of **7** to **9** depend on the naturally-occurring alkaloid sparteine as a source of chirality. Unfortunately, only one enantiomer of sparteine is available. Peter O'Brien of the University of York has developed (*J. Org. Chem.* **2004**, *69*, 5789) an alternative tricyclic amine **8** that is complementary to sparteine, directing oxidation toward *R*-**7**.

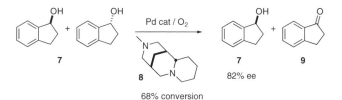

Asymmetric Nucleophilic Epoxidation

November 22, 2004

Asymmetric epoxidation of a prochiral alkene is an appealing process because two stereogenic centers are established in the course of the reaction. Often, the starting alkene is inexpensive. There have been several interesting recent advances in the asymmetric nucleophilic epoxidation.

Keiji Maruoka of Kyoyo University reports (*J. Am. Chem. Soc.* **2004**, *126*, 6844) the development of enantiomerically-pure quaternary ammonium salts (**2**) that catalyze the epoxidation of enones. The epoxidation of the *t*-butyl ketone **1** is particularly interesting, as Baeyer-Villiger oxidation would be expected to convert **3** into the ester **4**.

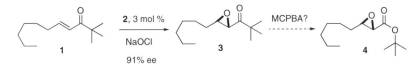

Masakatsu Shibasaki of the University of Tokyo reports (*J. Am. Chem. Soc.* **2004**, *126*, 7559) that use of a BINOL-derived catalyst with cumyl hydroperoxide enables the enantioselective epoxidation of unsaturated N-acyl pyrroles such as **7**. The pyrroles **7**, prepared from the precursor aldehydes such as **5** with the reagent **6**, can be used directly, without further purification.

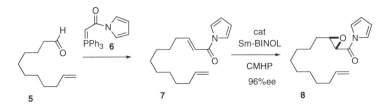

The product epoxy pyrroles such as **8** can be efficiently converted to the alcohol **9**, the homologated ester **10**, and the homologated ketone **11**.

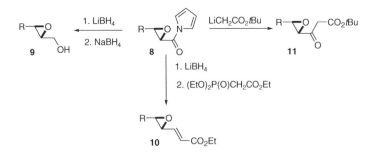

The Juliá-Colonna epoxidation uses poly-L-leucine and hydrogen peroxide to effect enantioselective epoxidation of chalcone derivatives such as **12**. In a pair of back-to-back papers (*Tetrahedron Lett.* **2004**, *45*, 5065 and 5069), H.-Christian Militzer of Bayer HealthCare AG, Wuppertal, reports a detailed optimization of this procedure. In the following paper (*Tetrahedron Lett.* **2004**, *45*, 5073), Stanley Roberts of the University of Liverpool reports the extension of this procedure to unsaturated sulfones such as **14**.

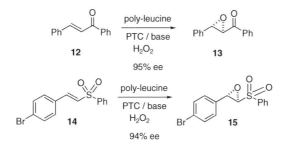

Asymmetric Synthesis of Nitrogen Heterocycles

November 29, 2004

The asymmetric synthesis of nitrogen heterocycles is of continuing interest to medicinal medicinal and bio-organic chemistry. Gareth Rowlands of the University of Sussex reports (*Tetrahedron Lett.* **2004**, *45*, 5347) diastereoselective routes to nitrogen heterocycles from the aziridine starting material **1**. Cu-mediated cyclization of **2** gives the bicyclic amine **3**, and Cu-mediated addition of **4** to **1** followed by *in situ* cyclization gives the lactam **5**.

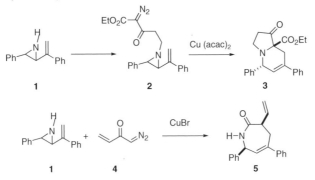

The reaction chemistry of aziridines is underdeveloped. This is particularly surprising because aziridines are easily prepared in high enantiomeric purity. Arlette Solladié-Cavallo of the Université L. Pasteur, Strasbourg, reports (*J. Org. Chem.* **2004**, *69*, 1409) that addition of the pulegone-derived sulfonium ylide **6** to an aldehyde tosylimine such as **7** proceeds to give the N-tosyl aziridine **8** in high ee and with good diastereocontrol. The sulfide precursor to **6** is recovered in almost quantitiative yield. N-Tosyl aziridines such as **8** are readily opened both by carbon and by heteroatom nucleophiles.

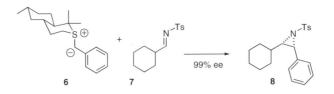

Amino acids continue to be useful starting materials for the preparation of enantiomerically-pure heterocycles. Henk Hiemstra of the University of Amsterdam and Floris Rutjes of the University of Nijmegen report (*J. Am. Chem. Soc.* **2004**, *126*, 4100) that cyclization of the allyl silane **9** followed by ring-closing metathesis leads to the highly-functionalized quinolizidine **11**.

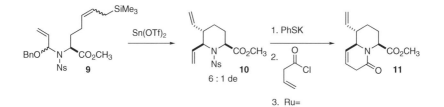

Synthesis of Amphidinolide T1

December 6, 2004

The amphidinolides are a class of structurally diverse and physiologically potent natural products. The key step in the total synthesis of enantiomerically-pure amphidinolide T1 **3** recently reported (*J. Am. Chem. Soc.* **2004**, *126*, 998) by Timothy Jamison of MIT, the Ni-mediated cyclization of **1** to **2**, clearly illustrates the power of organometallic C-C bond formation in organic synthesis.

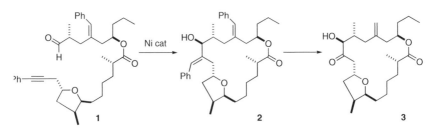

The alcohol fragment of **1** was prepared by Evans alkylation of **4** to give, after reduction and protection, the alkyne **5.** Ni-mediated coupling of the alkyne of **5** with the enantiomerically-pure epoxide **6**, following the procedures developed by the Jamison group, led to the alcohol **7** with high regio- and geometric control.

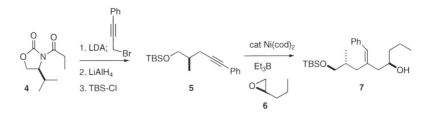

The acid portion of **1** was assembled by enantio-and diastereocontrolled addition of Z-crotyl borane to the aldehyde **8**, following the Brown protocol. Hydroboration and oxidation led to **9**, which was condensed with the allenyl silane **10** to give **11** with high diastereocontrol. Conversion of the alcohol to the iodide followed by three-carbon homologation by the Myers procedure then led to **1**, which was cyclized with > 10:1 regio- and diastereocontrol to give **12**. Ozonolysis and methylenation of the less hindered ketone then delivered **3**.

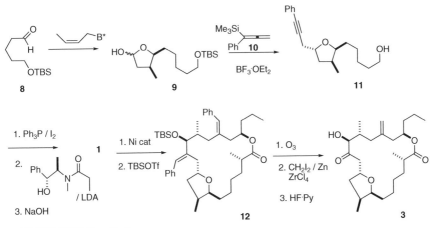

In both of the Ni-mediated steps in this synthesis, the Ni-alkyne complex is acting as an acyl anion, in one case opening an epoxide and in the other case adding to the aldehyde in an intramolecular sense. Such Ni-reduced phenylalkynes are among the easiest to prepare and least expensive of acyl anion equivalents.

Enantioselective C-C Bond Construction: Part One of Two

December 13, 2004

Enantioselective target-directed synthesis, which is important both for single-enantiomer pharmaceuticals and for natural product total synthesis, depends on the ability to form carbon-carbon bonds with absolute stereocontrol. An ideal method would be readily available, easy to practice, and proceed with high stereocontrol. In this column and the next one, we will review the most prominent recent developments.

The addition of an allylic organometallic can proceed to give mixtures both of regioisomers, and of geometric isomers. Junzo Nokami of the Okayama University of Science reports (*Organic Lett.* **2004**, *6*, 1261) that addition of an allylic organometallic to the inexpensive isomenthone **1** proceeds to give mainly the easily-separated branched adducts **2**. One of those adducts will, in the presence of acid, transfer the allyl group to an aldehyde with >99% ee and, remarkably, with > 99% *Z* geometric control. Such transfers have been reported before, with, for instance, menthone, but always to give the *E* product.

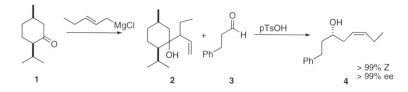

The enantioselective α-allylation of a ketone has become increasingly important. Martin Hiersmann of the Technische Universität Dresden reports (*Tetrahedron Lett.* **2004**, *45*, 3647) that α-ketoesters readily form enol ethers, such as **6**. On exposure to an enantiomerically-pure Cu catalyst, the enol ethers undergo facile Claisen rearrangement, leading to the allylated ketone (e.g. **7**) with high enantiomeric excess.

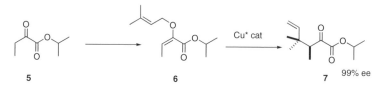

An alternative allylic coupling has been reported (*J. Am. Chem. Soc.* **2004**, *126*, 10676 and 11130) by Amir Hoveda of Boston College. Coupling of the allylic phosphonate **8** with diethyl zinc in the presence of an enantiomerically-pure Cu catalyst gives the branched product **9** in 95% ee. It is striking that the reaction works almost as well with the allylic phosphonate **10**. Reaction with diethyl zinc in the presence of the enantiomerically-pure Cu catalyst forms the quaternary center of **11** in 91% ee.

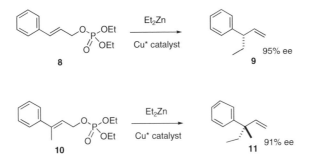

Enantioselective C-C Bond Construction:
Part Two of Two

December 20, 2004

Enantioselecitve target-directed synthesis, which is important both for single-enantiomer pharmaceuticals and for natural product total synthesis, depends on the ability to form carbon-carbon bonds with absolute stereocontrol. An ideal method would be readily available, easy to practice, and proceed with high stereocontrol. In this column, we conclude our review of the most prominent recent developments.

Asymmetric conjugate addition is a powerful method for the construction of ternary stereogenic centers. Erick Carreria of the ETH-Hönggerberg, Zürich reports (*Organic Lett.* **2004**, *6*, 2281) the chiral-auxiliary mediated conjugate addition of alkynes to alkylidene malonate derivatives such as **1**, to give, after hydrolysis and decarboxylation, the enantiomerically-enriched acid **3**.

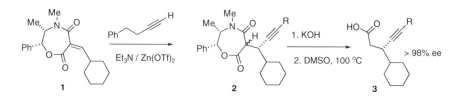

Enantioselective conjugate addition can also be carried out with cyclic enones. Shuichi Oi and Yoshio Inoue of Tohoku University in Sendai report (*Tetrahedron Lett.* **2004**, *45*, 5051) that a BINAP complex of Rh catalyzes the enantioselective conjugate addition of alkenyl Zr species such as **5**, to give **7** in high enantiomeric excess. Alkenyl Zr species such as **5** are readily prepared by hydridozirconation of alkynes. It is particularly important that addition of TMS-Cl to the reaction mixture at the end of the conjugate addition leads cleanly to the enol ether **6**.

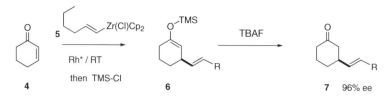

The asymmetric Strecker reaction, the addition of cyanide to an aldehyde imine, is one of the better ways to prepare α-amino acids. Masakatsu Shibasaki and Motomu Kanai of the University of Tokyo report (*Tetrahedron Lett.* **2004**, *45*, 3147, 3153) that use of a catalyst prepared from Gd isopropoxide and a glucose-derived phosphine oxide allows the asymmetric Strecker reaction to be carried out on *ketone*-derived imines such as **8**. Aryl alkyl ketones work well, as do alkyl methyl ketones. It is especially noteworthy that imines such as **8** derived from long-chain α,β-unsaturated ketones also give addition with high ee.

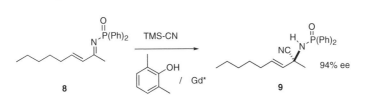

Advances in Nitrogen Protection and Deprotection

December 27, 2004

Nitrogen has a central role in medicinally-oriented organic synthesis. As it may not be convenient to carry an unprotected amine through several steps of a synthesis, N protection and deprotection becomes important.

The hydrolysis of a hindered amide can often be difficult. V. Bavetsias of the Cancer Research UK Laboratory in Surrey reports (*Tetrahedron Lett.* **2004**, *45*, 5643) that methanolic $Fe(NO_3)_3 \cdot 9H_2O$ will smoothly hydrolyze pivalamides such as **1** at room temperature. If this proves to be a general procedure for hindered amides, it will be a welcome addition to the armamentarium of organic synthesis.

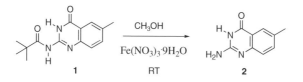

Usually, amides **7** are prepared from acids **6** and amines. C. Gürtler of Bayer MaterialScience AG in Leverkusen reports (*Tetrahedron Lett.* **2004**, *45*, 2515) the develop of catalysts for the alternative condensation of an acid **6** with an isocyanate **4**. It is particularly exciting that isocyanates are intermediates in the one-carbon degradation of an acid **3** to the corresponding amine. Current practice, if the *protected* amine were desired, is that the intermediate isocyanate **4** would be trapped with an alcohol, leading to the urethane **5**. This newly-reported observation offers the alternative of ending with the amide **7**, or perhaps with the sulfonamide **9**.

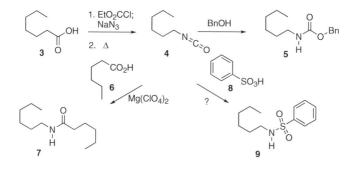

For amine protection, sulfonamides such as **9** offer several advantages over urethanes **5** or amides **7**. In particular, secondary amines protected as the urethane or the amide exist as mixtures of rotational isomers, confusing NMR characterization and making crystallization more difficult. The limitation has been that sulfonamides have been difficult to remove. Masanobu Uchiyama of the University of Tokyo reports (*J. Am. Chem. Soc.* **2004**, *126*, 8755) the development of transition metal ate complexes that catalyze electron-transfer reduction. While the sulfonamide **10** is inert to Mg in THF, inclusion of a catalytic amount of the ate complex **11** led to **12** in quantitative yield.

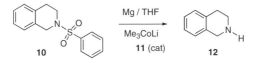

Enantioselective Synthesis of (+)-Tricycloavulone

January 3, 2005

Triclavulone **3**, recently isolated from *Clavularia vulgaris*, is one of the most complex of the family of about forty structurally-related fatty acid-derived marine prostanoids described from this species. Hisanaka Ito and Kazuo Iguchi of the Tokyo University of Pharmacy and Life Science recently reported (*J. Am. Chem. Soc.* **2004**, *126*, 4520) the total synthesis of **3**, starting with the preparation of the enantiomerically-enriched bicyclic ketone **1**.

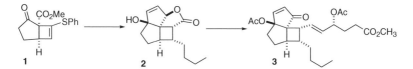

The enantioselective Cu-catalyzed cycloaddition of the reactive enone **4** to the alkyne **5** proceeded efficiently, but with only modest enantiomeric excess. This was improved at a later point in the synthesis, by recrystallization of **2**.

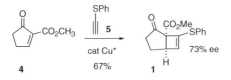

The folded geometry of **1** directed the Grignard addition, to give, after protection, the ester **7**. Homologation to the allylic alcohol **8** set the stage for Grubbs ring closure, to give, after oxidation, the tricyclic sulfone **9**, having the skeleton of triclavulone **3**.

There were two more stereocenters to set. It was expected that cuprates would add to the open face of the strained cyclobutene. The control of the other stereocenter was more problematic. One solution was to prepare an α-sulfonyl lactone. To this end, the ketone was converted to the secondary carbonate. As hoped, conjugate addition was followed by intramolecular acylation, but the reaction continued to full acyl transfer, to give **10**. Fortunately, desilylation of **10** proceeded with concomitant lactonization. Desulfonylation then gave **2**, which could be brought to high ee by recrystallization.

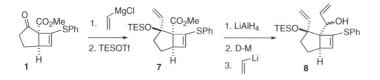

With **2** in hand, the rest of the synthesis proceeded smoothly. Reprotection followed by reduction and oxidation gave the keto aldehyde **11**. Condensation of the keto phosphonate **12** with **11** gave the enone **13**. Enantioselective transfer hydrogenation of **13** gave the allylic alcohol **14** with 11:1 diastereoselectivity. Protecting group interchange then gave triclavulone **3**, identical in every respect with natural material.

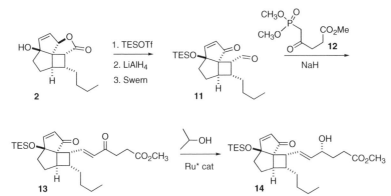

Best Methods of C-C Bond Formation:
Part One of Three

January 10, 2005

Carbon-carbon bond formation is fundamental to all of organic chemistry. Nucleophilic displacement is still the basis for most of what we do, but over the past thirty years radical addition and organometallic coupling have both been brought to a level of practical importance.

The conventional wisdom had been that since organometallic coupling necessitates formation of a carbon-transition metal bond, it is necessarily limited to substrates that lack β-hydrogens, since β-hydrde elimination would be faster than intermolecular C-C bond formation. In recent years, systems have developed for the effective coupling of *primary* halides having β-hydrogens. Greg Fu at MIT has now reported (*J. Am. Chem. Soc.* **2004**, *126*, 1340) the development of ligand systems that allow the efficient cross-coupling of *secondary* halides such as **1** having β-hydrogens.

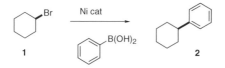

Professor Fu has also developed (*J. Am. Chem. Soc.* **2004**, *126*, 82) *ligand-free* methods for the effective Pd-mediated coupling of sp^2-hybridized organozirconium reagents with primary halides. This procedure is compatible with a wide range of organic functional groups. The organozirconium reagents are prepared directly from the corresponding terminal alkynes by the addition of the commercially-available Cp_2ZrClH. The ability to carry out the coupling without supporting ligands obviates the needs to remove the ligand after the reaction.

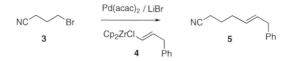

The difficulty of removing the last traces of the catalytic transition metal, especially of Pd, has put limits on the use of transition metal coupling reactions in the late stages of active pharmaceutical preparation. Kyotomi Kaneeda recently reported (*J. Am. Chem. Soc.* **2004**, *126*, 1604) the development of dendrimer-supported Pd complexes that efficiently catalyze Heck coupling. Solvent partitioning efficiently separated the Pd dendrimer from the coupled product. Since the products in this case were hydrocarbon-soluble, a thermomorphic hexane-DMF mixture was used for the separation. What will be needed for pharmaceutical preparation will be a dendrimer that is designed to partition the other way, into the hexane, leaving the pharmaceutical product in the DMF.

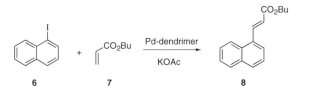

6   7   8

Best Methods of C-C Bond Formation:
Part Two of Three

January 17, 2005

Carbon-carbon bond formation is fundamental to all of organic chemistry. The emphasis this week is on recently-developed useful transformations that are easily scalable. Notably, these transformations telescope two or more synthesis steps into one-pot procedures.

Usually, to homologate an alcohol such as **1** to the corresponding nitrile **2** one would expect to first convert the alcohol into a leaving group. Nasser Iranpoor and Habib Firouzabadi of Shiraz University, Iran, have shown (*J. Org. Chem.* **2004**, *69*, 2562) that on exposure to a combination of Ph_3P, Bu_4NCN and DDQ, **1** is converted directly to **2**. The stereochemical outcome was not mentioned, but one would expect the reaction to proceed with net inversion, as illustrated.

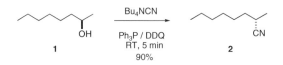

Often, esters are homologated by one carbon using the diazomethane-based Arndt-Eistert procedure. Kowalski homologation, addition of the inexpensive dibromomethane followed by α-elimination, is a more scalable alternative. Timothy Gallagher of the University of Bristol recently reported (*J. Org. Chem.* **2004**, *69*, 4849) the use of Kowalski homologation to prepare β-amino esters from α-amino esters, including the conversion of **3** to **4**. Note that the transformation can be carried out without protection of the OH, and that it proceeds without loss of stereochemical integrity.

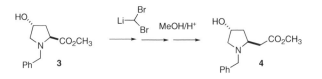

A third scalable method for one-carbon homologation, the conversion of an ester **5** to the terminal alkyne **7**, was reported (*Tetrahedron Lett.* **2004**, *45*, 5597) by Kevin Hinkle of GlaxoSmithKline in Research Triangle Park, NC. The ester is reduced with Dibal, the intermediate is quenched with methanol, and the solution of liberated aldehyde is treated with K_2CO_3 and the reagent **6** according to the Ohira procedure. Weinreb esters also work well with this protocol. The conversion was reported to work in 72% yield starting with 27 g of **5**. Note that **6** has a DSC exotherm at about 70°C, so caution should be exercised in handling it.

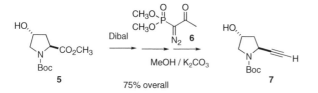

75% overall

Usually, one would think of alkylating a ketone such as **8** with an alkyl halide to homologate it to **10**. Often, aldol condensation with an aldehyde will proceed in higher yield, but this then requires three steps, aldol addition, dehydration, and hydrogenation, to reach **10**. Now Kiyotomi Kaneda of Osaka University has described (*J. Am. Chem. Soc.* **2004**, *126*, 5662) a Ru-grafted hydrotalcite catalyst that will mediate *four* steps, oxidation of **9** to the aldehyde, aldol condensation, dehydration and hydrogenation, in a single pot. The homologation also works with alcohols and arylacetonitriles.

In Memoriam: We note the untimely passing in December 2004 of Dr. Conrad Kowalski, a world-class contributor to new methods development and to scientific administration. We will miss both his chemical insight and his ready wit.

Best Methods of C-C Bond Formation:
Part Three of Three

January 24, 2005

Carbon-carbon bond formation is fundamental to all of organic chemistry. The emphasis again this week is on practical, scalable methods.

Formylation of organometallics is usually carried out with dimethylformamide (DMF). T. Ross Kelly of Boston College found (*J. Org. Chem.* **2004**, *69*, 2191) that although *o*-lithiation of **1** proceeded smoothly, formylation with DMF and with methyl formate failed. The inexpensive Fe(CO)$_5$, however, worked smoothly.

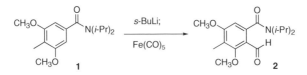

The Wittig reaction can be controlled to give a high percentage of the cis alkene. Young Gyu Kim of Seoul National University has shown (*Tetrahedron Lett.* **2004**, *45*, 3925) that addition of an excess of methanol to the –78°C Wittig reaction of **3** led to the complementary trans alkene **4**.

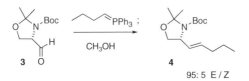

Another way to construct alkenes is by the addition of carbon radicals to nitrostyrenes such as **5**. Ching-Fa Yao of National Taiwan Normal University in Taipei has reported (*J. Org. Chem.* **2004**, *69*, 3961) an extension of this work, generating the acyl radical from the aldehyde **6**, cyclizing it to generate a new radical, then trapping that radical with **5** to give **7**. This article includes an overview of the several ways of adding radicals to **5**.

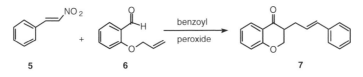

Telescoping of reaction steps, where feasible, can significantly improve the overall efficiency of a target-directed sequence. Zoapatanol **10** features a delicate combination of organic functional groups, in particular the β,γ-unsaturated ketone. In the last step of a total synthesis of **10** (*Organic Lett.* **2004**, *6*, 2149), Janine Cossy of ESPCI, Paris, describes the addition of prenyl lithium to the Weinreb amide **8**. The intermediate **9** was stable to dissolving metal reduction, where the ketone **10** would not have been. Work up then gave zoapatanol **10**. In the preceding paper (*Organic Lett.* **2004**, *6*, 2145), the authors describe in detail the development of this method. Note that in natural zoapatanol **10**, the stereogenic center adjacent to the ketone is a 1:1 epimeric mixture.

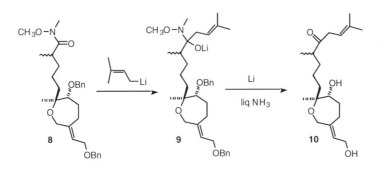

Formation of Aromatic-Amino and Aromatic-Carbon Bonds

January 31, 2005

The formation of bonds to aromatic rings is the foundation of much of organic synthesis. There has been much excitement over the past several years around Pd- or Cu-mediated displacement of an aromatic halide or aryl sulfonate with an amine. Paul Knochel of the University of Munich reports (*Angew. Chem. Int. Ed.* **2004**, *43*, 897) a complementary approach, the addition of an aryl halide-derived Grignard such as **2** to the diazonium derivative **3**. Aryl triflates such as **4** are efficient partners for further coupling.

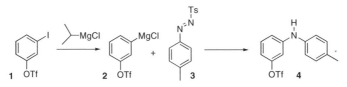

Although triflates such as **4** participate efficiently in coupling reactions, they are expensive. The attention of several groups has been focused on the development of less expensive alternatives. Qiao-Sheng Hu of CUNY Staten Island recently reported (*J. Am. Chem. Soc.* **2004**, *126*, 3058) that arene tosylates and benzenesulfonates such as **5** can serve as efficient leaving groups for Ni-catalyzed coupling to an areneboronic acid, to give the biaryl **7**. While tosylates are commonly used, benzenesulfonyl chloride has advantages on scale, since it is a liquid and so can be metered into a reaction.

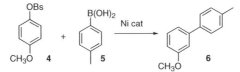

Carbon-carbon bond formation *ortho* to a directing group has a long history in organic synthesis. Tsuyoshi Satoh of the Tokyo University of Science has reported (*Tetrahedron Lett.* **2004**, *45*, 5785) a new protocol, based on the reaction of a deprotonated aniline such as **8** with a chloro sulfoxide such as **7**. An N-methyl aniline works equally well. The reaction may be proceeding by way of a coordinated alkylidene carbene.

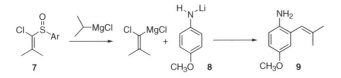

Supercritical CO_2 offers many advantages over conventional solvents for organic reactions. The challenge has been to find surfactants that will allow typical organic substrates to dissolve. Shu Kobayashi of the University of Tokyo has described (*J. Org. Chem.* **2004**, *69*, 680) the development of perfluoroalkyl aromatics that serve well, as illustrated by the conversion of **10** to **11**. Both the perfluoroalkyl aromatic and the Sc(OTf)$_3$ are recovered at the end of the reaction.

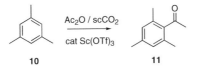

Synthesis of Dendrobatid Alkaloid 251F

February 7, 2005

The Dendrobatid "poison arrow" frogs of Central and South America exude a potent mixture of alkaloids from their skins. It was originally thought that the frogs biosynthesized these alkaloids, but it has since been shown that they are of dietary origin. The skin exudate of the Colombian frog *Minyobates bombetes* causes severe locomotor difficulties, muscle spasms and convulsions upon injection in mice. The major component of the alkaloid mixture is 251F **3**. Jeff Aubé of the University of Kansas recently described (*J. Am. Chem. Soc.* **2004**, *126*, 5475) the enantioselective total synthesis of **3**. The key step in the synthesis was the cyclization of the keto azide **2**.

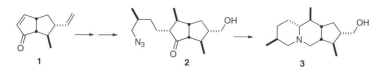

The ketone **1**, with four ternary stereogenic centers in one cyclopentane ring, was a significant challenge for synthesis. A clever solution flowed from the idea of coupling a chiral Diels-Alder reaction with alkene metathesis. The relative and absolute configuration of **1** were set by the Diels-Alder cycloaddition of the acyl oxazolidine **4** to cyclopentadiene, to give **5**. Homologation of **5** to the enone **6** set the stage for alkene metathesis, mediated by the Grubbs catalyst **7**, to give **1**.

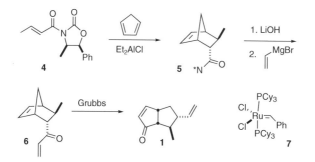

Conjugate addition to **1** proceeded across the open face of the bicyclic system to give an enolate, condensation of which with the enantiomerically-pure aldehyde **8** gave the enone **9**. Conjugate reduction of the enone also removed the benzyl ether, to give the alcohol. Conversion of the alcohol to the azide gave **10**. Ozonolysis followed by selective reduction then gave **2**.

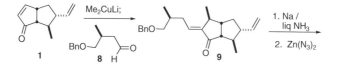

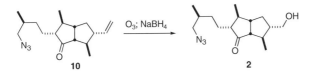

The acid-mediated intramolecular addition of the azide to the ketone proceeds by way of addition followed by a pinacol-type rearrangement, to give the amide **12**. The regioselectivity of the bond migration is remarkable. Reduction of the amide then gave **3**.

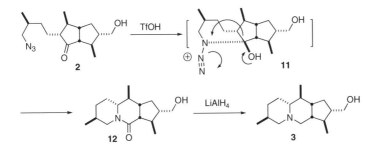

Enantioselective Construction of Aldol Products: Part One of Two

February 14, 2005

Acyclic stereoarrays are important both in themselves and as precursors to enantiomerically-defined ring systems. Although the aldol condensation has long been a workhorse for acyclic stereoselection, there are still new things being done.

An exciting development in recent years has been the use of small organic molecules as catalysts for asymmetric transformations. Several years ago, it was reported that proline would catalyze the enantioselective addition of acetone to aldehydes. Following up on this observation, Yun-Dong Wu of the Chengdu Institute of Organic Chemistry reports (*J. Am. Chem. Soc.* **2003**, *125*, 5262) the development of the proline amide **2** as a catalyst for this condensation. The catalyst **2** gave consistently higher ee's than did proline.

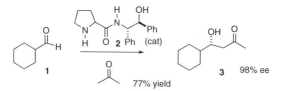

The aldol condensation can also be used to construct quaternary stereogenic centers. James L. Gleason of McGill University reports (*Organic Lett.* **2004**, *6*, 405) that reduction of the sulfide **4** with dissolving metal gives a lithium enolate. Conversion to the boron enolate followed by addition to benzaldehyde gave the product **5** in high diastereomeric excess. The authors ascribe the observed high stereocontrol to geometric control in the formation of the intermediate enolate.

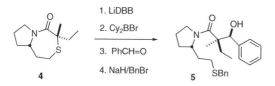

"Aldol" products do not have to come from an aldol condensation. In another example of catalysis by a small organic molecule, Jeffrey Bode of UC Santa Barbara reports (*J. Am. Chem. Soc.* **2004**, *126*, 8126) that the thioazolium salt **7** effects the rearrangement of an epoxy aldehyde such as **6** to the aldol product **8**. This is a net oxidation of the aldehyde, and reduction of the epoxide. As epoxy aldehydes such as **6** are readily available by Sharpless asymmetric epoxidation, this should be a general route to enantiomerically-aldol products. The rearrangement also works with an aziridine aldehyde such as **9**, to give the β-amino ester **10**.

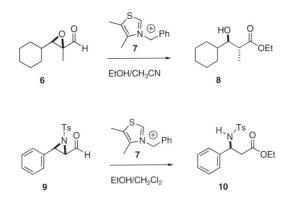

Enantioselective Construction of Aldol Products: Part Two of Two

February 21, 2005

   Acyclic stereoarrays are important both in themselves and as precursors to enantiomerically-defined ring systems. Although the aldol condensation has for many years been a workhorse for acyclic stereoselection, there are still new things that can be done.

   In a pair of papers last year, Scott Nelson of the University of Pittsburgh expanded the range of the ketene "aldol". In the first paper (*J. Am. Chem. Soc.* **2004**, *126*, 14), he employed a chiral Al-based catalyst **3**. This catalyst mediated additions such as propionyl bromide **1** to **2** to give **4** in 98:2 syn/anti ratio and 95% ee.

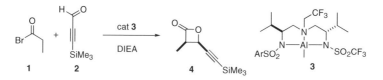

   In a follow-up paper (*J. Am. Chem. Soc.* **2004**, *126*, 5352), Professor Nelson used the commercially-available base quinaldine **7a** (R=H) or its TMS ether **7b** (R=TMS). Catalysts **7a** and **7b** are both efficient and give > 95% ee, but lead to opposite absolute configurations of the products. As with catalyst **3**, the syn aldol product predominates, but now *branched* aldehydes such as **6** also participate efficiently in the reaction. This is another example of enantioselective catalysis by a small organic molecule.

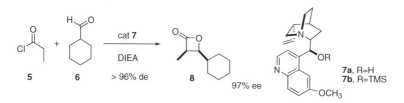

   Mukund Sibi of North Dakota State University has developed (*J. Am. Chem. Soc.* **2004**, *126*, 718) a powerful three-component coupling, combining an α,β-unsaturated amide **9**, a hydroxylamine **10**, and an aldehyde **11**. The hydroxylamine condenses with the aldehyde to give the nitrone, which then adds in a dipolar sense to the unsaturated ester. The reaction proceeds with high diastereocontrol, and the absolute configuration is set by the chiral Cu catalyst. As the amide **9** can be prepared by condensation of a phosphonacetate with another aldehyde, the product **12** can be seen as the product of a four-component coupling, chirally-controlled aldol addition and Mannich condensation on a starting acetamide.

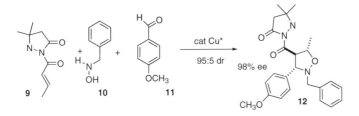

Enantioselective α-Functionalization of Carbonyl Compounds

February 29, 2005

The enantioselective oxygenation procedures, epoxidation and dihydroxylation, developed by Barry Sharpless have dominated single-enantiomer organic synthesis. Recently, several additional methods for enantioselective oxidation have been developed, based on the α-functionalization of carbonyl compounds.

This area is active enough that in two instances the development was reported essentially simultaneously by two different research groups. Hisashi Yamamoto of the University of Chicago (*J. Am. Chem. Soc.* **2004**, *126*, 5360) and Armando Córdova of Stockholm University (*Angew. Chem. Int. Ed.* **2004**, *43*, 1109) independently reported the enantioselective α-oxygenation of ketones. Professor Yamamoto employed Ag-BINAP **3** to catalyze the condensation of a ketone-derived enol ether with nitrosobenzene **2**. The trimethylstannyl enol ether **1** led *in THF* to the α-oxygenated product **4** in high enantiomeric excess. The tributylstannyl enol ether **5** *in DME* gave the complementary α-aminated product **6**, again in excellent ee. Professor Córdova produced **4**, also in >99% ee, by condensing cyclohexanone itself with **2**, with (*S*)-proline as the catalyst in CHCl₃ or DMSO.

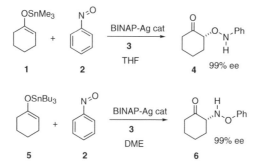

With cyclic enones, Professor Yamamoto has developed (*J. Am. Chem. Soc.* **2004**, *126*, 5962) an enantioselective double functionalization. The organocatalyst **8** mediates conjugate addition of N and α´-oxygenation, to give **9**.

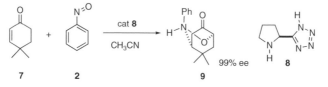

Enantioselective halogenation is a powerful transformation, directly installing an efficient leaving group. Thomas Leckta of Johns Hopkins University has shown (*J. Am. Chem. Soc.* **2004**, *126*, 4245) that benzoylquinine **11** catalyzes the α-chlorination of ketenes derived from acid chlorides such as **10**, to give **12** in high ee.

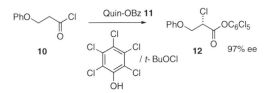

In Communications submitted two weeks apart, David W.C. MacMillan of Caltech (*J. Am. Chem. Soc.* **2004**, *126*, 4108) and Karl Anker Jorgensen of Aarhus University, Denmark (*J. Am. Chem. Soc.* **2004**, *126*, 4790) reported the enantioselective α-chlorination of aldehydes, using organocatalysts **14** and **15** respectively. The chloro aldehydes are promising precursors to, *inter alia*, enantiomerically-pure epoxides.

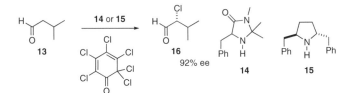

Synthesis of (-)-Hamigeran B

March 7, 2005

Of all the ring-forming reactions of organic synthesis, diastereoselective intramolecular Diels-Alder reactions are among the most powerful. Often, as illustrated by the cyclization of **1** to **2**, a single stereogenic center can set the relative and absolute configuration of two rings. The cyclization of **1** to **2** is the key step in the total synthesis of (-)-hamigeran B recently reported (*J. Am. Chem. Soc.* **2004**, *126*, 613) by K.C. Nicolaou of the Scripps Research Institute.

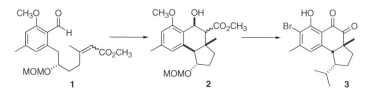

The preparation of **1** started with the addition of lithiated **4** to the enantiomerically-pure epoxide **5**, which was prepared from the racemate using the Jacobsen protocol. Reduction followed by selective protection of the primary alcohol gave the monosilyl ether, which was further protected with MOM chloride to give **7**. Pd-mediated oxidation to the methyl ketone followed by condensation with the Horner-Emmons reagent gave the unsaturated ester **8** as an inconsequential mixture of geometric isomers. Oxidation then set the stage for the crucial cyclization.

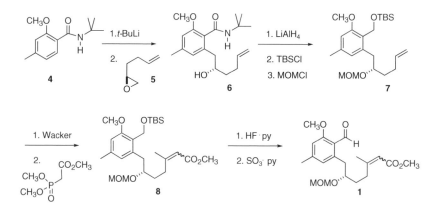

On irradiation, the aldehyde **1** underwent photoenolization to give the quinone methide **9**. Intramolecular Diels-Alder cyclization then proceeded with high diastereocontrol to give **2** as a mixture of epimeric esters.

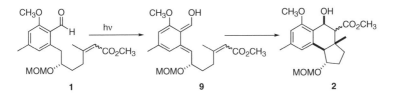

1   9   2

The ester **2** has a trans 6-5 ring fusion, whereas in the desired **3** the ring fusion is cis. This was easily corrected, as the 6-5 ring fusion is more stable cis. Acid-mediated dehydration to give the alkene proceeded with concomitant removal of the MOM protection. Osmylation followed by acetonide formation and subsequent oxidation gave the ketone, which was readily epimerized to **10**. The acetonide was designed to block H addition to the bottom side, so hydrogenation of **11** would give the isopropyl group endo. In fact, hydrogenation led to the undesired exo isopropyl, but hydroboration proceeded from the exo face, leading to the desired **12**.

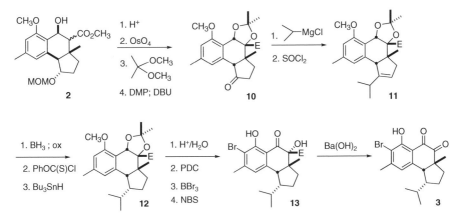

Hydrolysis of the acetonide followed by oxidation and bromination provided the ketone **13**, which is itself a natural product, hamigeran A. Hydrolysis under aerobic conditions led first to decarboxylation, then to autooxidation, to give (-)-hamigeran B **3**.

Catalytic C-C Bond Forming Reactions

March 14, 2005

Usually, carbon-carbon bonds are formed by coupling two carbons each of which are already functionalized in some way, as with the displacement of a C-Br with NaCN to form C–CN. It would be more efficient, and potentially less expensive and less polluting, if one of the partners could be an ordinary C-H bond. *Intramolecular* processes for carbene insertion into unactivated C-H bonds have been known for years. Practical *intermolecular* processes for C-C bond formation to a C-H bond are just starting to appear.

There are two ways to approach this problem. One way is to activate the target C-H with the organometallic catalyst. This process can be remarkably selective. Yasutaka Ishii of Kansai University, Osaka has reported (*J. Org. Chem.* **2004**, *69*, 1221) that Pd-mediated oxidation of **1**, for instance, proceeds predominantly at the para position, leading, after Heck coupling, to the product **2**.

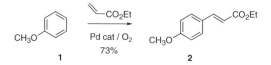

The C-H bond does not have to be attached to an aromatic ring. Robert Bergman and Jonathan Ellman of the University of California at Berkeley have reported (*Organic Lett.* **2004**, *6*, 1685) that a Rh catalyst will activate **4** to add to simple alkenes such as **3** in a reductive sense, to give **5**.

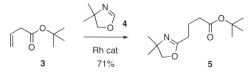

The other way to effect C-H activation is to add a carbene or carbenoid to the C-H bond. One of the most promising ways to generate the required carbene is by catalytic rearrangement of an alkyne. Chuan He of the University of Chicago has observed (*J. Org. Chem.* **2004**, *69*, 3669) that an Au catalyst will rearrange ethyl propiolate **7** to give an intermediate that inserts into aromatic C-H bonds to give the Z-alkene **8**.

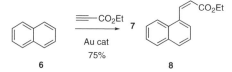

Insertion can also be carried out on the C-H bonds of heteroaromatics. Masahiro Murakami of Kyoto University has described (*J. Am. Chem. Soc.* **2003**, *125*, 4720) a Ru catalyst that will effect rearrangement of a silyl alkyne such as **10** into the vinylidene carbene. The intermediate Ru carbene complex is then electrophilic enough to insert into the aromatic C-H bond. The insertion is highly regioselective. The Au and the Ru alkylidene insertions are geometrically complementary, as Ru gives the E-alkene.

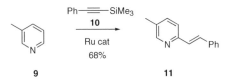

Rare Sugars Are Now Readily Available Chriral Pool Starting Materials

March 21, 2005

Many of the five- and six-carbon sugars, although well known, are rare, and too expensive to be used as chiral pool starting materials. David MacMillan of Caltech in an elegant series of papers (*Angew. Chem. Int. Ed.* **2004**, *43*, 2152; Science **2004**, *305*, 1752; *Angew. Chem. Int. Ed.* **2004**, *43*, 6722) has demonstrated a two-step route not just to protected six-carbon carbohydrates, but also to alkyl, thio and amino derivatives of those carbohydrates.

The key observation was that L-proline would catalyze the addition of α-hetero aldehydes to α-branched aldehydes such as **2** to give the aldol product **3** with high enantio- and diastereocontrol. Even more exciting, in the absence of other acceptors the α-hetero aldehydes *dimerize* with high relative and absolute stereocontrol. Both alkoxy and silyloxy aldehydes worked efficiently.

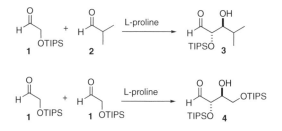

The second step in the hexose synthesis was the aldol condensation of **4** with another α-hetero aldehyde derivative **5**. By tuning the Lewis acid and the solvent, three of the four possible diastereomeric products could be selectively prepared. α-Amino and α-thio aldehydes worked well also, leading to **9** and **10** respectively.

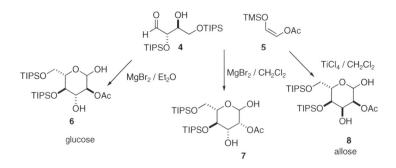

Proline catalysis leads to the anti products **3** and **4**. Use of the designed imidazolidinone catalyst **11** leads to the complementary syn product **12**.

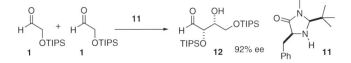

Alkyne Metathesis in Organic Synthesis

March 28, 2005

While alkyne metathesis is not going to displace alkene methathesis as a synthetic method, it is a complementary approach that can offer advantages.

Catalysts for alkyne metathesis are still under active development. A recent paper by Karol Grela of the Polish Academy of Sciences in Warsaw (*J. Org. Chem.* **2004**, *69*, 7748) provides an detailed overview of the options. The choice is between sensitive preformed catalysts that provide high turnover but require more exacting organometallic techniques, or, alternatively, in situ catalysts that require higher temperature and longer reaction times, but are less expensive and less technically challenging to prepare. For the latter, the inexpensive $Mo(CO)_6$ has been the precursor of choice, with an added phenol ligand. In this paper, Professor Grela and co-workers optimized the supporting phenol, finding that 2-fluorophenol was the most effective. Dimerizations with this catalyst system require no special precautions – indeed, they can be run open to the air. Both cyclodimerization (**1** →**2**) and cross metathesis (**3**→**4**) proceeded efficiently.

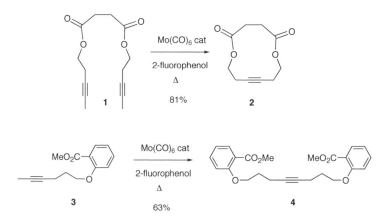

The more widely used alkene metathesis is deficient in that the alkenes so prepared are often mixtures of geometric isomers. Alkyne metathesis offered what promised to be a general solution to this problem. Hydrogenation of the alkyne products to the Z-alkene was straightforward. Reduction of the isolated alkyne to the E-alkene was not so obvious. Alois Fürstner of the Max-Planck-Institut, Mülheim, has surveyed (*Tetrahedron* **2004**, *60*, 7315) several approaches, then optimized the most promising, Ru-mediated hydrosilylation followed by protodesilylation. The procedure worked equally well to construct E,E-dienes. This promises to be a mild and general solution to the long-standing challenge of E-reduction of an alkyne in the presence of other functional groups.

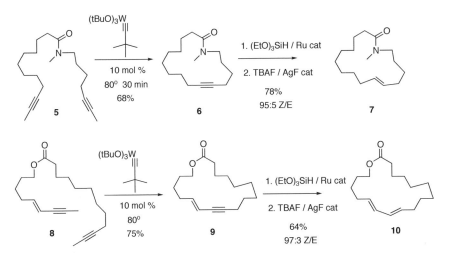

In each of the alkyne metatheses outlined here, the byproduct is the volatile 2-butyne. The alkyne metathesis can only be carried out on internal alkynes, since the metathesis catalysts cyclotrimerize terminal alkynes such as **11** to benzene derivatives. In this context, it may prove useful that readily-available terminal alkynes such as **14** are easily isomerized specifically to methyl alkynes such as **15** (*Tetrahedron Lett.* **1990**, *31*, 5843).

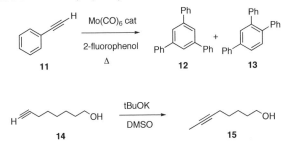

Total Synthesis of (±)-Sordarcin

April 4, 2005

Carbohydrate derivatives of sordaricin **3** are clinically-effective antifungal agents, but development efforts were halted when a suffficient supply of **3** could not be established. Koichi Narasaka of the University of Tokyo recently reported (*Chem. Lett.* 2004, 33, 942) a total synthesis of **3**, based on the elegant Pd-mediated cyclization of **1** to **2**.

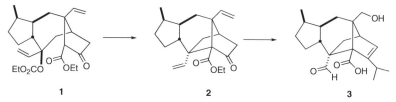

1   **2**   **3**

The 5-7-5 skeleton of **1** was assembled from cyclohexenone **4** by conjugate addition followed by enolate trapping. Simmons-Smith cyclopropanation of **5** led to the cyclopropyl alcohol **6**. Generation of the oxy radical from **6** led to fragemntation to give **7**, which further cyclized to give **8**. Note that the seven-membered ring is flexible enough that the radical cyclization delivers the required trans ring fusion. Annulation to **9** followed by kinetically-controlled conjugate addition then gave **10**.

The B-keto ester was protected as the enol acetate, then the ketone **12** was homologated to the

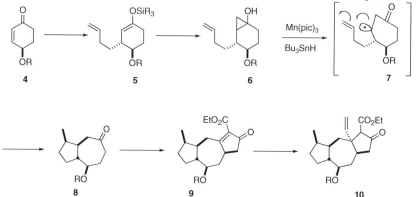

8   **9**   **10**

carbonate **1**. Despite the strain in the [2.2.1] system being formed, the Pd-mediated cyclization of **1** to **2** proceeded smoothly.

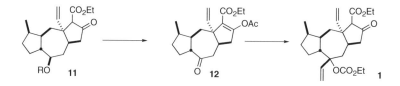

11   **12**   **1**

128

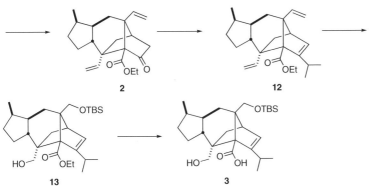

2

12

13

3

To complete the synthesis, the ketone of **2** was homologated to the alkene **12**. Selective oxidative cleavage of the two vinyl groups followed by reduction provided the diol, the less encumbered alcohol of which was protected to deliver the fully-differentiated ester **13**. Oxidation followed by ester cleavage then gave **3**.

Ru-Mediated Intramolecular Alkene Metathesis: Improved Substrate and Catalyst Design

April 1, 2005

As alkene metathesis is applied to more and more challenging substrates, difficulties have arisen, that in turn have driven creative solutions. One such is the relay ring-forming metathesis strategy put forward (*J. Am. Chem. Soc.* **2004**, *126*, 10210) by Thomas Hoye of the University of Minnesota. For metathesis to proceed, the catalyst must first interact with one of the two alkenes. If neither is reactive, the metathesis will not work. Professor Hoye conceived the idea that a pendant reactive alkene could interact quickly with the catalyst, then lose a cyclic alkene to generate the requisite intermediate carbene complex. In addition to its utility for cyclizing unreactive dienes, this strategy can be used to steer metathesis when more than one option is available. Cyclization of **6**, for instance, delivers nearly equal amounts of **3** and **4**. Cyclization of **1**, on the other hand, delivers predominantly **4**, while cyclization of **5** gives **3**. Note that selectivity is higher with the first-generation Grubbs catalyst than with the second generation catalyst.

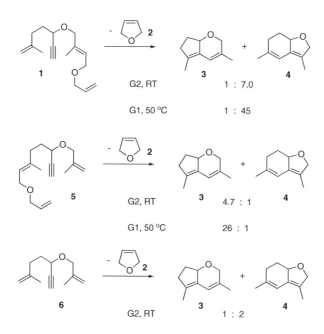

This strategy has already been found useful in natural product synthesis. In the course of a synthesis of V-ATPase inhibitor oximidine III, John Porco of Boston University has described (*Angew. Chem. Int. Ed.* **2004**, *43*, 3601) the cyclization of **7** to **8**. In the absence of the pententyl director, the initial complexation of the Ru catalyst was with the 1,3-diene, leading to allylidene complex and so effectively killing the catalyst. In this case, the Hoveyda catalyst **8** provided a cleaner product than G2 did.

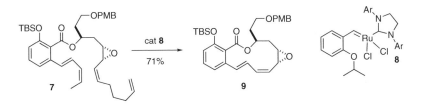

The usual metathesis catalysts G1, G2 and **8** exist in the reaction medium primarily in the resting form, so 5-10 mol% of catalyst is often required to drive reactions to completion. Warren Piers of the University of Calgary has prepared (*Angew. Chem. Int. Ed.* **2004**, *43*, 6161) a rapidly-initiating Ru catalyst **12** that does not have such a resting state. This catalyst, which is stable to ambient air and moisture and can be maintained in refluxing CD_2Cl_2 for several hours with no sign of decomposition, is effective at loadings of 0.1 mol %.

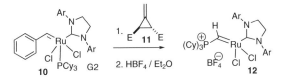

Heterocyclic Construction by Grubbs Metathesis

April 18, 2005

Grubbs metathesis has proven to be a powerful tool for the rapid construction of complex heterocycles. Shengming Ma of the the the Shanghai Institute of Organic Chemistry reports (*J. Org.* Chem. **2004**, *69*, 6305) a concise assembly of the tetracycle **4**. Exposure of the symmetrical triene **1** to the second-generation Grubbs catalyst led first to ene-yne metathesis, then to diene metathesis, to give the triene **2**. Addition of maleimide **3** proceeded with substantial diastereocontrol, to give **4**.

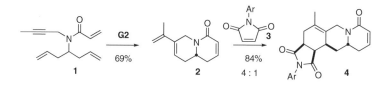

Many macrocyclic lactams have potent physiological activity. Daesung Lee of the University of Wisconsin has taken advantage (*Organic* Lett. **2004**, *6*, 4351) of the conformational preference of diacyl hydrazides such as **8** to prepare, by Grubbs cyclization of **8** and then again of **10**, the 8-8 system **11**. Exposure of **11** to Na in liquid ammonia reduced the N-N bond and opened the epoxide, to deliver macrocyclic lactam **12**.

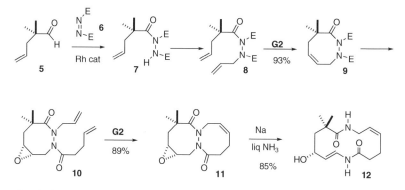

Faced with the preparation of BILN 2061 **18**, a hepatitis C NS3 protease inhibitor, Anne-Marie Faucher of Boehringer Ingelheim (Canada) in Quebec adopted (*Organic Lett.* **2004**, *6*, 2901) a ring-closing metathesis approach. To this end, it was necessary to prepare the enantiomerically-enriched amino acid **14**. Remarkably, Knowles hydrogenation of **13** proceeded efficiently, without concomitant reduction of the very reactive monosubstituted alkene. For the cyclization of **15** to **17**, the Hoveyda metathesis catalyst **16** proved the most efficient. Note that although a large ring is being formed, it is sufficiently constrained that only the desired Z-alkene **17** is observed.

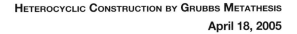

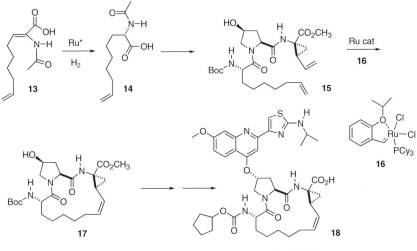

BILN 2061

Natural Product Synthesis Using Grubb Metathesis: Lasubine II, Ingenol, and Ophirin B

April 25, 2005

Practitioners of total synthesis have been pushing the limits of Grubbs metathesis. Siegfried Blechert of the Technisches Universität, Berlin, envisioned (*Tetrahedron* **2004**, *60*, 9629) that Grubbs metathesis of **1** could open the cyclopentene, to give a new Ru alkylidene that could condense with a styrene such as **2**. In practice, this transformation worked well, yielding **3**. Deprotection, intramolecular Michael addition and reduction then gave (-)-lasubine II **4**.

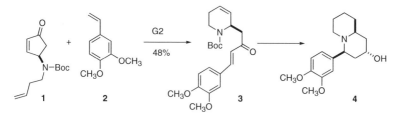

The twisted inside-outside skeleton of ingenol has long challenged organic chemists. Hideo Kigoshi of the University of Tsukuba conceived (*J. Org. Chem.* **2004**, *69*, 7802) that metathesis of **6** could lead to **7**, leading directly into an established ingenol end game. The diene **6** was readily prepared by intramolecular alkylation of **5**, with bond formation occurring away from the adjacent methyl group. As expected. Alkylation with methallyl iodide then proceeded exo on the cis-fused bicyclic skeleton, to give the desired **6**. The key question was whether or not the two alkenes of **6** could find each other in the metathesis reaction. In fact, this proceeded smoothly. Alkene **7** likely does not react with the Grubbs catalyst, so once the ring forms, it cannot reverse.

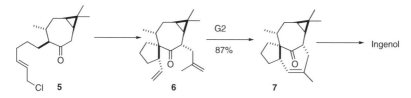

There are times that it is important to run the Grubbs reaction under *equilibrating* conditions. In the course of a synthesis of ophirin B **12**, Michael Crimmins of the University of North Carolina observed (*J. Am. Chem. Soc.* **2004**, *126*, 10264) that metathesis of **8** under the usual conditions gave mainly the undesired cyclic dimer. At elevated temperature, the dimer re-entered the equilibrium, leading to the desired **9** as the major (15:1) product. The oxonene ring system of **10** then directed the intramolecular Diels-Alder cycloaddition, leading to **12**.

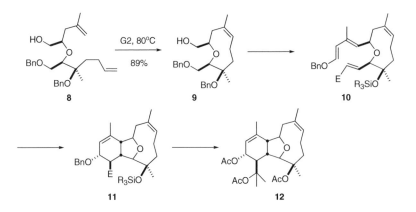

Synthesis of (-)-Tetrodotoxin

May 2, 2005

Tetrodotoxin **3**, the toxic principle of pufferfish poison, is a formidable challenge for synthesis, with each carbon of the cyclohexane functionalized. Minoru Isobe of Nagoya University recently reported (*Angew. Chem. Int. Ed.* **2004**, *43*, 4782) a second-generation synthesis of enantiomerically-pure tetrodotoxin. This syntheis features the rapid construction of the cyclohexene by Diels-Alder cycloaddition using an enantiomerically-pure dienopile, the early introduction of the aminated quaternary center, and the use of that center to direct the relative configuration of further functionalization around the ring.

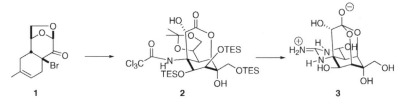

The synthesis started with levoglucosenone **4**, available by the pyrolysis of cellulose, e.g. old newspapers. Bromination-dehydrobromination gave the enantiomerically-pure Diels-Alder dienophile **5**, which was combined with isoprene to give predominantly the crystalline adduct **1**. Hydrolysis and acetylation led to **6**, which was carried on to the geometrically-defined allylic alcohol **7** via reduction with Zn-Cu couple. Overman rearrangement of **7** proceeded with high facial control, to give **8**.

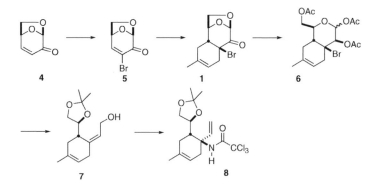

The next stage of the synthesis was ring oxygenation, to convert **8** into **13**. The key to this transformation was the observation that the amide oxygen of **8** participated in the solvolysis of the allylic bromide, setting, after hydrolysis, the new secondary stereocenter of **9**. Hydroxyl-directed epoxidation gave **10**, which was rearranged with Ti(O-i-C$_3$H$_7$)$_4$ to **11**. After some experimentation, it was found that the derived dione **12** could be reduced to the desired cis diol **13** with LiBr and LiAlH(O-t-C$_4$H$_9$)$_3$ followed by NaBH$_4$/CeCl$_3$.

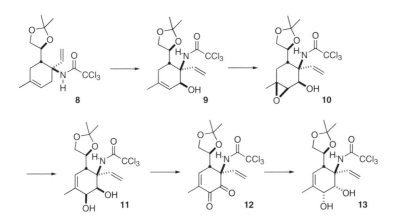

Silylation followed by selenium dioxide oxidation converted **13** into **14**. Epoxidation of the derived TES ether proceeded by addition of oxygen to the more open face of the alkene, leading to **15**. Ozonolysis followed by diastereoselective one-carbon homologation provided **17**. This set the stage for intramolecular epoxide opening by the carboxylate, to give **2**, in which all of the stereogenic centers of tetrodotoxin have been established.

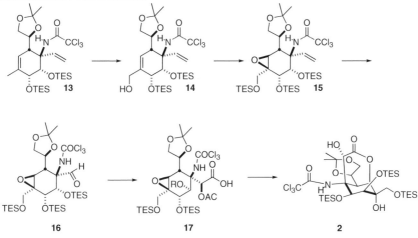

Justin Du Bois of Stanford University has put forward (*J. Am. Chem. Soc.* **2003**, *125*, 11510) a quite different total synthesis of tetrodotoxin, including an elegant late-stage introduction of the nitrogen.

Diastereoselective and Enantioselective Construction of Aza-Heterocycles

May 9, 2005

Five- and six-membered azacyclic rings are the basic building blocks of pharmaceutical synthesis. Several powerful methods for the construction of such rings that recently have been developed are highlighted here.

Continuing the theme of small molecules as catalysts for organic reactions, Eric Jacobsen of Harvard has reported (*J. Am. Chem. Soc.* **2004**, *126*, 10558) the design of a peptide thiourea that mediates enantioselective Pictet-Spengler cyclization, e.g. of **1** to **2**.

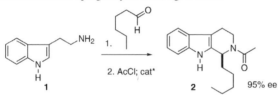

Günter Helmchen of the Universität Heidelberg took advantage (*Chem. Comm.* **2004**, 896) of the substitutional flexibility of π-allyl iridium complexes to develop enantioselective cyclizations such as **3** to **4**. Six-membered rings are also formed efficiently (88% ee).

In a related approach, John Wolfe of the University of Michigan has demonstrated (*Angew. Chem. Int.Ed.* **2004**, *43*, 3605) that Heck addition of an aryl halide **5** to an amino alkene **6** can be terminated by C-N bond formation. The reaction proceeds with high diastereoselectivity, to give **7**.

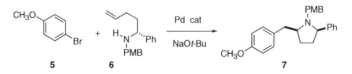

Azacycles can also be constructed by C-C bond construction. The enantiomerically-pure amide **8** is easily prepared by photolysis of pyridine followed by acetylation and enzymatic desymmetrization.

Patrick Mariano of the University of New Mexico has shown (*J. Org. Chem.* **2004**, *69*, 7284) that the second generation Grubbs Ru catalyst converts the derived **9** into **10** with high regioselectivity.

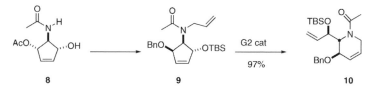

Jeff Aubé of the University of Kansas has reported (*Org. Lett.* **2004**, *6*, 4993) a cascade strategy of C-C and C-N bond formation. Thus, Diels-Alder cyclization of **11** and **12** gives **13**, which in situ undergoes azido-Schmidt ring expansion to give **14**. In a complementary approach, **15** and **16** combine to give **17** and then **18**.

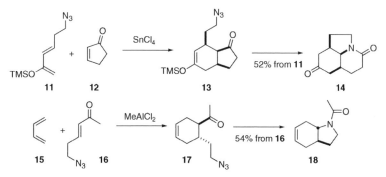

There is a continuing interest in the diastereoselective and enantioselective hydrogenation of inexpensive pyridine derivatives. Frank Glorius of the Max-Planck-Institut, Mülheim, has coupled (*Angew. Chem. Int. Ed.* **2004**, *43*, 2850) 2-chloropyridines such as **19** with the inexpensive chiral auxiliary **20**. Hydrogenation of **21** then proceeded with high diastereocontrol, to give the reduced product **22** and recovered **20**.

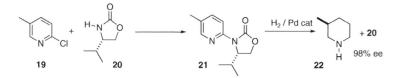

New catalyst systems for intramolecular alkene hydroamination have also been developed (*Chem. Comm.* **2004**, 894 and *Angew. Chem. Int. Ed.* **2004**, *43*, 5542).

Diastereoselective and Enantioselective Construction of Cyclic Ethers

May 16, 2005

Because the stereocontrolled construction of cyclic ethers has been difficult and expensive, the use of cyclic ethers as pharmaceuticals has not been fully explored. With the development of powerful new methods for the diastereoselective and enantioselective construction of cyclic ethers, this situation is changing. The best of the recently developed methods for stereocontrolled cyclic ether construction are highlighted here.

Often, ethers are constructed by alkoxide displacement of a stereodefined leaving group. Hirakazu Arimoto of Nagoya University has found (*Chem. Commun.* **2004**, 1220) that it is possible to effect diastereocontrolled construction of cyclic benzylic ethers such as **2** by oxidation of **1** to the o-quinone methide. The more stable equatorial C-glycoside product **2** is formed in near quantitative yield.

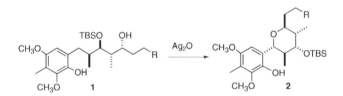

Triply-convergent "linchpin" construction is a powerful approach to organic synthesis. Tarun Sarkar of the Indian Institute of Technology, Kharagpur reports (*Angew. Chem. Int. Ed.* **2004**, *43*, 1417) the preparation of the valuable bis-silane **5**. Condensation of **5** with **3** and **4** proceeded in a highly diastereoselective fashion, to give the five-membered ring ether **6**.

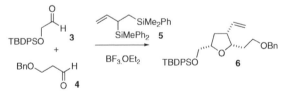

Distal C-C bond formation can also be used to construct cyclic ethers, but this demands that methods be developed for the enantioselective assembly of complex acyclic ethers. P. Andrew Evans of Indiana University has demonstrated (*Angew. Chem. Int. Ed.* **2004**, *43*, 4788) that Rh-mediated coupling of secondary allylic carbonates such as **7** with secondary alcohols such as **8**, both enantiomerically pure, proceeds with clean retention (double inversion) of absolute configuration. Alkene metathesis then delivers the cyclic ether **9** in high diastereomeric and enantiomeric purity.

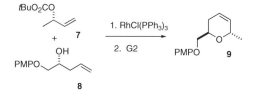

We (*J. Org. Chem.* **2004**, *69*, 7234) used the power of the Sharpless oxidations to convert the prochiral **10** into the epoxy diol **11**. Base-catalyzed cascade cyclization then converted **11** into crystalline **12**, again with high diastereomeric and enantiomeric purity. An advantage of this approach is that by changing the absolute sense of the epoxidation and/or the dihydroxylation, it should be possible to selectively prepare each of the four enantiomerically-pure diastereomers of **12**.

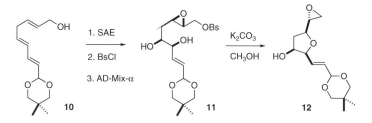

Amir Hoveyda of Boston College has developed (*J. Am. Chem. Soc.* **2004**, *126*, 12288) an elegant series of Ru catalysts for *enantioselective* alkene metathesis. The power of this approach is illustrated by the direct conversion of the easily-prepared prochiral alkene **13** into the enantiomerically-enriched cyclic ether **15**.

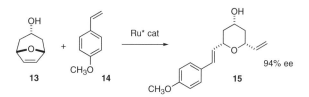

Synthesis of Heterocyclic Natural Products: (-)-Ephedradine A, (-)-α-Tocopherol, (-)-Lepadin D, and (-)-Phenserine

May 26, 2005

Stereocontrolled syntheses of *macro*lides and *macro*lactams are well developed. Much remains to be done toward the efficient enantioselective construction of *five* and *six*-membered cyclic ethers and amines. Four recent natural product syntheses illustrate the current state of the art.

In the first synthesis of the hypotensive alkaloid (-)-ephedradine A **3** (*Tetrahedron* **2004**, *60*, 9615), Tohru Fukuyama of the University of Tokyo faced the challenging of constructing the central five-membered ring ether with control of relative and absolute configuration. While neither chiral Rh catalysts nor chiral auxiliaries alone gave satisfactory results, a combination of the two worked efficiently, yielding the two trans diastereomers of **2** in a 13:1 ratio.

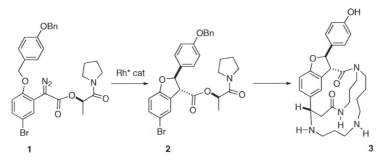

Vitamin E is a collective term for all of the tocopherols and tocotrienols. Of these, one of the most active is (-)-α-tocopherol **6**. Lutz Tietze of the Universität Göttingen has reported (*Angew. Chem. Int. Ed.* **2005**, *44*, 257) that the Pd-catalyzed cascade cyclization of **4** to **5** proceeds with 96% ee.

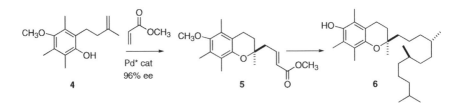

Dawei Ma of the Shanghai Institute of Organic Chemistry (*Angew. Chem. Int. Ed.* **2004**, *43*, 4222) set the absolute configuration of the lepadins, exemplified by (-)-lepadin D **11**, using a chiral pool starting material. Condensation of the bromide **8**, prepared from L-alanine, with 1,3-cyclohexanedione **7** gave the enamide **9**. Hydrogenation of **9** proceeded with high diastereocontrol, to give the ketone **10**, which was then carried on to several members of the lepadin family.

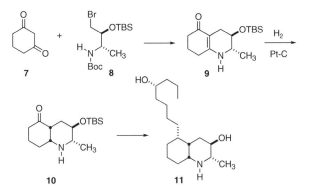

Larry Overman also used (*J. Am. Chem. Soc.* **2004**, *126*, 14043) a chiral pool starting material, but in a different way. The prochiral enolate **12** showed substantial diastereoselectivity in its reaction with the bis-triflate **13**, almost 10:1. Through the power of algebra, it followed that the three diastereomers of **14** were formed in a ratio of 90 : 9 : 1. The crystalline **14** was easily isolated in diastereomerically-pure form, and carried on to phenserine **15**. This is a new method for the stereocontrolled construction of chiral quaternary centers.

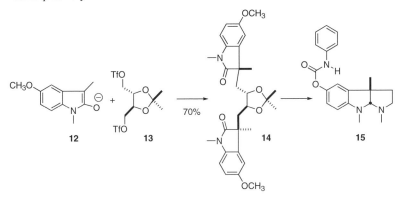

Protection of N- and O-Functional Groups

May 30, 2005

O- and N-protection is often necessary in organic synthesis. Several recent advances in functional group protection-deprotection are particularly noteworthy.

Primary benzenesufonamides have been notoriously difficult to protect. Theodore S. Widlanski of Indiana University has now found (*Tetrahedron Lett.* **2004**, *45*, 8483) that while N-benzyl sulfonamides such as **1** are resistant to hydrogenolytic debenzylation, the easily-prepared Boc derivative **2** is smoothly debenzylated. Brief exposure of **3** to trifluoroacetic acid then gives the primary sulfonamide **4**.

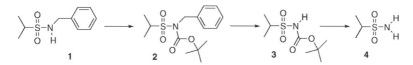

Amides are often cleaved with strong alkali. Fabio Prati of the Università di Modena has reported (*Organic Lett.* **2004**, *6*, 3885) that treatment of triphenyl phosphite with chlorine at –30 °C gives a substance that reacts smoothly with amides such as **5** to give the amine **6** as the HCl salt. The imino chloride is the intermediate, so this also provides a convenient entry to Bischler-Napieralski cyclization.

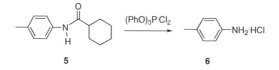

In addition to the more common allyl protecting groups, alcohols can be protected as methallyl and prenyl ethers. Pierre Vogel of the Swiss Federal Institute of Technology, Lausanne, has demonstrated (*Organic Lett.* **2004**, *6*, 2693) that using benzenesulfonyl radical, one can efficiently and selectively remove these protecting groups, one after the other.

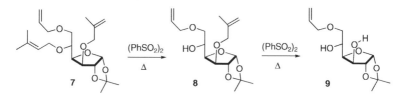

The MOM ether is a simple protecting group that would be more widely used if it were easier to put on. G.V.M. Sharma of IIT Hyderabad has found (*Tetrahedron Lett.* **2004**, *45*, 9229) that the inexpensive $ZrCl_4$ efficiently catalyzes the reaction of dimethoxymethane with an alcohol such as **10** to give the MOM ether **11**. MOM ethers are easily deprotected with the same catalyst in 2-propanol. Note that both reactions were carried out in the presence of other acid-labile functional groups.

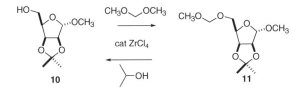

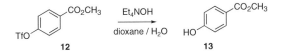

Phenol protection enjoys a special place in organic synthesis. In addition to being good leaving groups for coupling reactions, sulfonates, e.g. a triflate such as **12**, are attracting growing attention because of the way they change the reactivity around the arene ring. This necessitates eventual deprotection. Shigeru Nishiyama of Keio University, Yokohama has shown (*Tetrahedron Lett.* **2004**, *45*, 6317) that Et$_4$NOH in aqueous dioxane converts arene triflates into the corresponding phenols under mild conditions. Again, note that the easily-hydrolyzed methyl ester survives.

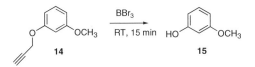

M.G. Finn of Scripps/La Jolla has explored (*Organic Lett.* **2004**, *6*, 2777) propargyl protection for phenols. The propargyl ether is removed by BBr$_3$ more easily than the usually-labile methyl ether. The propargyl group is removed from a primary alcohol under comparable conditions.

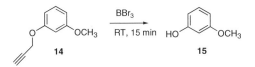

Synthesis of (-)-Norzoanthamine

June 6, 2005

The marine alkaloid (-)-norzoanthamine **3** suppresses bone loss in ovariectomized mice, and so is of interest as a lead to antiosteoporotic drugs. A substantial challenge in the assembly of **3** is the stereocontrolled construction of the C ring, with its three all-carbon quaternary centers. In the synthesis of **3** by Masaaki Mayashita of Hokkaido University (*Science* **2004**, *305*, 495), the B and C rings were built and two of the three needed quaternary centers were set by the intramolecular Diels-Alder cyclization of **1** to **2**.

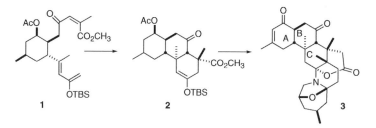

The triene **1** was prepared from the enone **4**, available in enantiomerically-pure form over several steps from pulegone. Triply-convergent coupling with the cuprate **5** and the aldehyde **6** led to the furan **7**. Functional group manipulation then gave **1**, setting the stage for the intramolecular Diels-Alder cyclization.

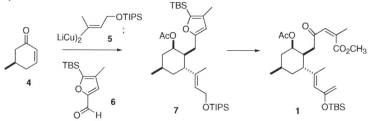

The cyclization of **1** proceeded with 72: 28 diastereoselectivity, leading, after hydrolysis, to the crystalline diketone **8** as the major product. Reduction of the ketones to the axial alcohols was followed by spontaneous lactonization, allowing easy differentiation of the several functional groups. Homologation to **10** followed by condensation with methyl carbonate and subsequent O-methylation then gave **11**. C-Methylation of **11** then set the third quaternary center of the C ring. The deuteriums were introduced to minimize an unwanted intramolecular hydride transfer in a later step.

Patrick Mariano of the University of New Mexico has shown (*J. Org. Chem.* **2004**, *69*, 7284) that the second generation Grubbs Ru catalyst converts the derived **9** into **10** with high regioselectivity.

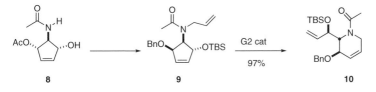

Jeff Aubé of the University of Kansas has reported (*Org. Lett.* **2004**, *6*, 4993) a cascade strategy of C-C and C-N bond formation. Thus, Diels-Alder cyclization of **11** and **12** gives **13**, which in situ undergoes azido-Schmidt ring expansion to give **14**. In a complementary approach, **15** and **16** combine to give **17** and then **18**.

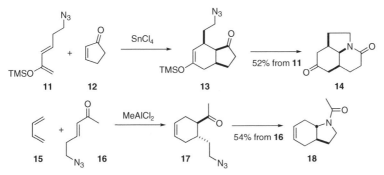

There is a continuing interest in the diastereoselective and enantioselective hydrogenation of inexpensive pyridine derivatives. Frank Glorius of the Max-Planck-Institut, Mülheim, has coupled (*Angew. Chem. Int. Ed.* **2004**, *43*, 2850) 2-chloropyridines such as **19** with the inexpensive chiral auxiliary **20**. Hydrogenation of **21** then proceeded with high diastereocontrol, to give the reduced product **22** and recovered **20**.

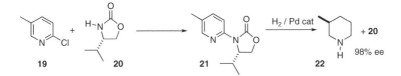

New catalyst systems for intramolecular alkene hydroamination have also been developed (*Chem. Comm.* **2004**, 894 and *Angew. Chem. Int. Ed.* **2004**, *43*, 5542).

Best Synthetic Methods:
C-C Bond Formation

June 13, 2005

Carbon-carbon bond formation is basic to organic synthesis. Progress is made by developing new transformations, but it also is made by developing more practical and scalable procedures for already-known transformations

Ketone methylenation is usually effected with $Ph_3P=CH_2$, scarcely an atom-efficient process. Tu-Hsin Yan of National Chung-Hsing University reports (*Organic Lett.* **2004**, *6*, 4961) that reduction of CH_2Cl_2 with Mg powder in the presence of $TICl_4$ leads to a reagent that efficiently homologates even easily-enolizable ketones such as **1**. The reagent also converts esters such as **3** to alkenes, albeit more slowly.

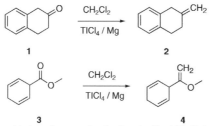

Carbon monoxide is an inexpensive feedstock. David J. Cole-Hamilton of St. Andrews University has found (*Chem. Comm.* **2004**, 1720) Pd catalysts that effect isomerization of internal alkenes. The transient less-stable terminal alkene is selectively homologated to the corresponding ester. Remarkably, conditions can be tuned such that the alkene **5** can be converted to either **6** or **7**.

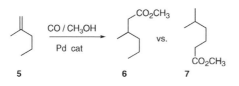

Carbonylation with CO in the presence of H_2 leads to aldehydes. Bernhard Breit of Albert-Ludwigs-Universität, Freiburg, has found (*Chem. Comm.* **2004**, 114) that conversion of an allylic alcohol to the *o*-diphenylphosphanylbenzoate **8** allows highly diastereoselective and regioselective Rh-mediated one-carbon homologation.

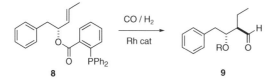

Nucleophilic organometallic reagents bearing functional groups are important intermediates in organic synthesis. David M. Hodgson of the University of Oxford has optimized (*Org. Lett.* **2004**, *6*, 4187) the metalation of a terminal epoxide, using *s*-BuLi and a designed diamine, DBB. The resulting anion adds efficiently to aldehydes, amides and Bu₃SnCl.

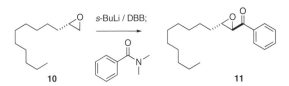

Paul Knochel of the Universität München has also been developing functionalized organometallics, using Grignard exchange on aromatic halides. He has now (*Angew. Chem. Int. Ed.* **2004**, *43*, 3333) extended his earlier work on aryl iodides to the less expensive aryl bromides such as **12**. Note that the organometallic reagent so produced will couple to even an unactivated primary iodide. Professor Knochel has also shown (*Org. Lett.* **2004**, *6*, 4215) that alkenyl iodides such as **14** exchange and couple under these conditions.

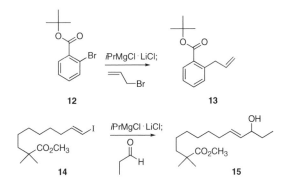

Enantioselective Construction of Single Stereogenic Centers

June 20, 2005

Single enantiomer synthesis is of increasing importance in pharmaceutical production. It is essential that practical and scalable procedures be developed for controlling the absolute configuration of new stereogenic centers as they are formed. Severally particularly important developments have recently been reported.

A challenge of long standing has been the enantioselective addition of a propargyl anion to an aldehyde. Teck-Peng Loh of the National University of Singapore has reported (*Chem. Commun.* **2004**, 2456) that the well-known allyl transfer from a defined secondary alcohol works well also with allenyl alcohols such as **1**. The reaction proceeds to give a product **3** of inverted absolute configuration compared to the donor. The enantiomerically-pure alcohol **1** was prepared by resolution, but Chan-Mo Yu of Sungkyunkwan University has described (*Chem. Commun.* **2004**, 2494) the preparation of alcohols such as **1** by the enantioselective reduction of the corresponding ketones.

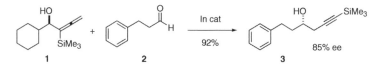

Erick Carreira of ETH Hönggerberg has reported (*Organic Lett.* **2004**, *6*, 4575) a catalytic procedure for the reduction of nitroalkenes such as **4** to the nitroalkane **5** with high enantiomeric excess. He has also reported (*Angew. Chem. Int. Ed.* **2005**, *44*, 612) that the enantiomerically-enriched nitroalkane can be converted to the corresponding nitrile **6** without loss of stereochemical purity.

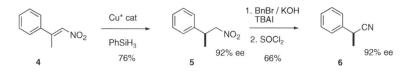

An alternative route to nitriles of high enantiomeric excess, based on the conjugate addition of Me_3Si-CN to an activated amide **7**, has been reported (*J. Am. Chem. Soc.* **2004**, *126*, 9928) by Eric Jacobsen of Harvard University. The key to success in this case was dual catalysis, with a chiral Al complex *and* a chiral Er complex.

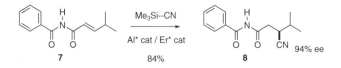

Another advance in conjugate addition has been reported (*Organic Lett.* **2004**, *6*, 4877) by Toshiro Harada of Kyoto Institute of Technology. The organocatalyst **10** efficiently mediates the addition of ketene silyl acetals to enones such as **9**. In this case, it was use of the *dimethylsilyl* group that led to success.

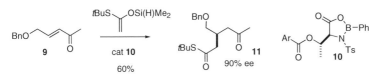

The enantioselective conjugate addition of *alkyl* groups to enones has required the use of dialkyl zinc reagents. Now, Ben Feringa of the University of Groningen has found (*J. Am. Chem. Soc.* **2004**, *126*, 12784) that with ferrocene-based chiral Cu catalysts, one can achieve conjugate addition with ordinary alkyl Grignard reagents.

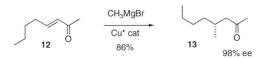

The work on organocatalytic aldol condensation continues unabated. In a recent advance, Armando Córdova of Stockholm University has found (*Angew. Chem. Int. Ed.* **2004**, *43*, 6528) that (*S*)-proline also will catalyze the *Mannich* reaction with high enantiomeric excess. Acyclic ketones also participate efficiently.

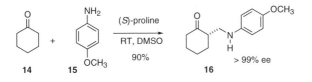

Enantioselective Construction of Arrays of Stereogenic Centers

June 27, 2005

Single enantiomer synthesis is of increasing importance in pharmaceutical production. It is essential that practical and scalable procedures be developed for controlling the absolute configuration of new stereogenic centers as they are formed. In the previous column, recent advances for the preparation of *single* stereogenic centers were covered. The construction of more extended arrays of stereogenic centers, which is also is important, is covered here.

One approach is to use enantioselective methods to establish a first stereogenic center, then use that center to control the relative configuration of additional centers as they are formed. Patrick Walsh of the University of Pennsylvania has found (*J. Am. Chem. Soc.* **2004**, *126*, 13608) that addition of a dialkyl zinc reagent to an aldehyde such as **1**, using the high e.e. Nugent procedure, gives an intermediate that on exposure to molecular oxygen gives the epoxy alcohol **2** with high diastereomeric control. In a complementary approach, Guofu Zhong of the Scripps Institute, La Jolla has shown (*Chem. Commun.* **2004**, 606) that enantioselective aminoxylation of an aldehyde such as **3** can be followed in the same pot by the addition of an organometallic reagent, to give the monoprotected diol **4** in high enantiomeric excess. While the diastereomeric control is low in this case, one would expect this to improve if a bridging Lewis acid were included.

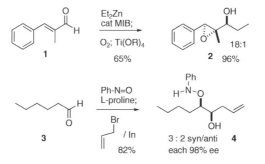

Enantioselective aldol reactions also can be used to create arrays of stereogenic centers. Two elegant α-amino anion approaches have recently been published. Fujie Tanaka and Carlos F. Barbas III of the Scripps Institute, La Jolla, have shown (*Org. Lett.* **2004**, *6*, 3541) that L-proline catalyzes the addition of the aldehyde **6** to other aldehydes with high enantio- and diastereocontrol. Keiji Maruoka of Kyoto University has developed (*J. Am. Chem. Soc.* **2004**, *126*, 9685) a chiral phase transfer catalyst that mediates the addition of the ester **9** to aldehydes, again with high enantio- and diastereocontrol.

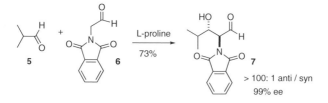

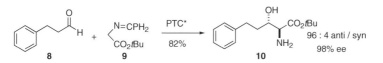

Michael addition can also be used to establish arrays of stereogenic centers. Hiyoshuzi Kotsuki of Kochi University has shown (*J. Am. Chem. Soc.* **2004**, *126*, 9558) that the chiral DMAP derivative **13** mediates the addition of cyclic ketones such as **11** to nitrostyrene **12** with high enantio- and diastereocontrol. Acyclic aldehydes also add with high stereocontrol. Li Deng of Brandeis University has developed (*Angew. Chem. Int. Ed.* **2005**, *44*, 105) a quinine-based catalyst **16** that directs the addition of **12** to a single face of the cyclic β-ketoester **15**, establishing adjacent ternary and quaternary centers. For the conversion of the nitro group to a nitrile without epimerization, see *Angew. Chem. Int. Ed.* **2005**, *44*, 612.

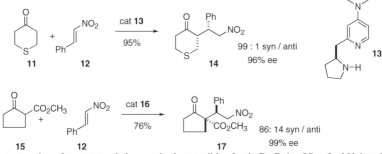

The construction of more extended arrays is also possible. Justin Du Bois of Stanford University has reported (*Angew. Chem. Int. Ed.* **2004**, *43*, 4349) diastereoselective remote functionalization using C-H insertion of a sulfamate such as **18** to give the oxathiazinane **19**, which on reaction with an allyl silane gives the alkylated product **20** with high diastereocontrol. Note that the oxygen of **20** is activated as a leaving group.

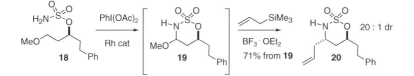

Synthesis of (+)-Brasilenyne

July 4, 2005

(+)-Brasilenyne **3**, isolated from the digestive gland of the sea hare *Aplysia brasiliana*, shows significant antifeedant activity. Scott Denmark of the University of Illinois has described (*J. Am. Chem. Soc.* **2004**, *126*, 12432) an elegant synthesis of **3**, the key step of which is the Pd-mediated intramolecular coupling of **1** to give **2**.

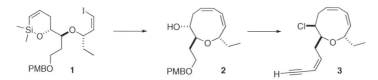

The enantiomerically-pure intermediate **1** was prepared from the dioxolanone **4**, available in three steps from L-malic acid. Lewis acid-mediated homologation converted **4**, a 4:1 mixture of diastereomers, into **5** as a *single* diastereomer. After establishment of the alkenyl iodide, it necessary to maintain the lactone in its open form. A solution was found in the formation of the Weinreb amide. The final stereogenic center was established by Brown allylation of the derived aldehyde. The alkene metathesis to form **1** was carried out with the commercially-available Schrock Mo catalyst. The authors did not comment on the relative efficacy of alternative alkene metathesis catalysts.

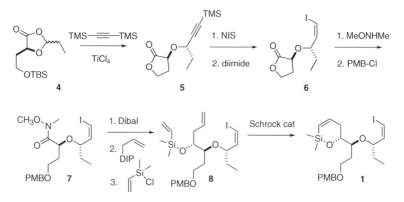

With **1** in hand, the stage was set for the proposed Pd-mediated coupling. The authors were pleased to observe that the coupling conditions previously developed in the group worked efficiently with this much more complex substrate, leading to **2** as a single geometric isomer. After protection-deprotection and oxidation, homologation using the Corey protocol gave **10**. Formation of the chloride proceeded with the expected clean inversion of absolute configuration, to give **3**.

With **1** in hand, the stage was set for the proposed Pd-mediated coupling. The authors were pleased to observe that the coupling conditions previously developed in the group worked efficiently with this much more complex substrate, leading to **2** as a single geometric isomer. After protection-deprotection and oxidation, homologation using the Corey protocol gave **10**. Formation of the chloride proceeded with the expected clean inversion of absolute configuration, to give **3**.

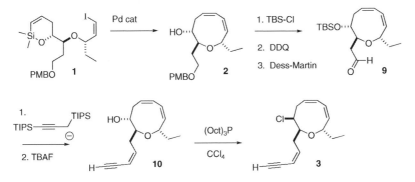

The least precedented transformation in this synthesis is the homologation of **4** to **5**. This appears to be a general solution to a longstanding challenge, the construction of secondary-secondary ethers with absolute stereocontrol.

Best Synthetic Methods: Functional Group Transformations

July 11, 2005

An advantage of polymer-based reagents is that both the excess and the spent reagent are easily separated from the product. Bruno Linclau of the University of Southampton has reported (*J. Org. Chem.* **2004**, *69*, 5897) the preparation of a polymer-bound carbodiimide. Exposure of the polymer to alcohol gives a family of *O*-alkylisoureas that smoothly convert carboxylic acids to the corresponding esters. Methyl, benzyl, allyl and *p*-nitrobenzyl transfer smoothly. The polymeric *t*-butyl reagent could not be prepared.

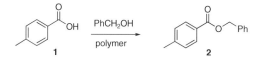

Many procedures have been developed for the conversion of an alcohol to the iodide. Nevertheless, the $ZrCl_4$-based procedure reported (*Tetrahedron Lett.* **2004**, *45*, 7451) by Habib Firouzabadi and Nasser Iranpoor of Shiraz University, Iran is promising. It works well with primary, secondary and tertiary alcohols under mild conditions, and there are no organic sideproducts.

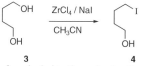

The conversion of an alcohol to the amine is often a multi-step procedure. Jonathan M.J. Williams of the University of Bath has described (*Chem. Commun.* **2004**, 1072) a direct Ir-catalyzed procedure for this transformation. The reaction probably involves oxidation of the alcohol to the aldehyde, imine formation, and then reduction of the imine to the amine. No secondary examples were reported.

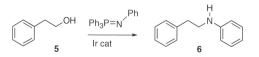

Oxidative decarboxylation of acids to alkenes is often accompanied by alkene rearrangement. Lukas J. Goossen of the Max-Planck-Institut, Mühlheim, has found (*Chem. Commun.* **2004**, 724) that *in situ* activation of the acid with phthalic anhydride and inclusion of the bis phosphine DPE-Phos substantially slow alkene isomerization, which can be essentially eliminated by running the reaction to only 80% conversion. Both linear and branched carboxylic acids work well.

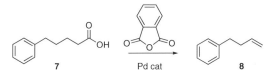

The specific construction of trisubstituted alkenes is a continuing challenge. Shigeru Nishiyama of Keio University has described (*Tetrahedron Lett.* **2004**, *45*, 8273) the specific bromination-dehydrobromination of the allylic ester **9**, delivering **10** as a 60:1 mixture of geometric isomers. Pd-mediated coupling led to the ester **11**.

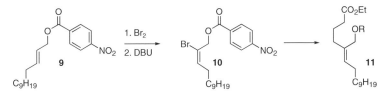

The least expensive starting organic "functional group" is a C-H bond. Two facile procedures for C-H activation have recently appeared. Melanie Sanford of the University of Michigan has developed (*J. Am. Chem. Soc.* **2004**, *126*, 9542) a procedure for specifically oxygenating methyl groups α to a ketone, by Pd-mediated oxidation of the derived methoxime **12**. The process is catalytic in Pd. John F. Hartwig of Yale University has found (*J. Am. Chem. Soc.* **2004**, *126*, 15334) that methyl groups can also be specifically functionalized by Rh-mediated oxidative borylation. In the case of **14**, steric effects direct the borylation to the ethyl group. The product borane **15** undergoes the expected oxidation and coupling reactions.

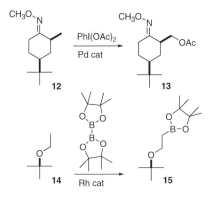

Enantioselective Construction of Oxygenated and Halogenated Secondary Stereocenters

July 18, 2005

High ee halogenated secondary centers and, via activation, oxygenated secondary centers are requisite intermediates for the preparation of enantiomerically-pure natural products and pharmaceuticals. Several methods have recently been reported for the conversion of achiral or prochiral starting materials into high ee intermediates.

A limitation on resolution is that the desired enantiomer is only half of the racemic starting material. Kurt Faber of the University of Graz has reported (*Org. Lett.* **2004**, *6*, 5009) a clever solution to this problem. On exposure of the sulfate **1** of a secondary alcohol to aerobically grown whole cells of *Sulfolobus acidocaldarius* DSM 639, one enantiomer of the sulfate was smoothly converted into the other enantiomer of the starting alcohol. The enzyme consumed the more reactive enantiomer > 200 times more rapidly than the less reactive enantiomer. For the last bit of conversion, the ee of the product alcohol will of course fall. One solution to this would be to run the reaction near 50% conversion, then hydrolyze the mixture to give high ee product alcohol **2**. Exposure of the mixture to a lipase that selectively acetylated the minor enantiomer would then polish the ee of **2**.

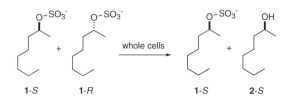

The ketone **3** is prochiral. Hisashi Yamamoto of the University of Chicago has shown (*J. Am. Chem. Soc.* **2004**, *126*, 15038) that conversion of a ketone such as **3** to its silyl enol ether followed by exposure to the chiral chlorinating agent **4** gives the chloro ketone **5** in high ee. The agent **4** is easily regenerated. Chloro ketones such as **5** can be reduced to the chloro alcohol with high diastereoselectivity.

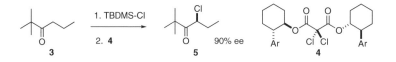

The epoxidation procedure developed by Yian Shi of Colorado State University has become one of the workhorses of enantioselective synthesis. That work has been based around trans and trisubstituted alkenes. Professor Shi has now developed (*Tetrahedron Lett.* **2004**, *45*, 8115) an efficient protocol for the enantioselective epoxidation of aryl-substituted *cis* alkenes such as **6**.

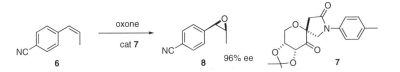

Masakatsu Shibasaki of the University of Tokyo has developed effective procedures for the epoxidation of α,β-unsaturated amides with high ee. He has now reported (*Angew. Chem. Int. Ed.* **2004**, *43*, 317) reagents for the selective reduction of the epoxy amide **9** to either the β-hydroxy amide **10** or the α-hydroxy amide **11**.

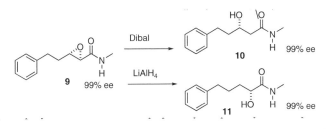

High ee epoxy alcohols such as **12** are prepared by Sharpless asymmetric epoxidation. Kohji Suda of Meiji Pharmaceutical University, Tokyo, has described (*J. Am. Chem. Soc.* **2004**, *126*, 9554) the optimization of Cr TPP as a catalyst for the rearrrangement of the epoxides such as **12** to the corresponding aldehyde. The reaction proceeds without loss of ee.

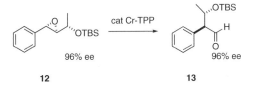

Enantioselective Construction of Aminated Secondary Stereocenters

July 25, 2005

The enantioselective construction of aminated stereogenic centers is a central task both for pharmaceutical production and for alkaloid synthesis.

Activated aziridines should be as useful as epoxides for carbon-carbon bond formation, with the advantage that the product will already incorporated the desired secondary aminated stereocenter. To date, a general enantioselective method for the aziridination of alkenes has not been developed. Eric Jacobsen of Harvard University (*Angew. Chem. Int. Ed.* **2004**, *43*, 3952) has explored an interim solution, based on the resolution of racemic epoxides such as **1**. The cobalt catalyst that selectively hydrolyzes one enantiomer of the epoxide also promotes the addition of the imide to the remaining enantiomerically-enriched epoxide. As expected, the aziridine **4** is opened smoothly with dialkyl cuprates.

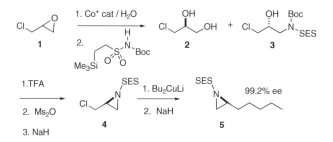

Two methods for the direct construction of enantiomerically-enriched allylic amines have recently been reported. John Hartwig of Yale University has developed (*Angew. Chem. Int. Ed.* **2004**, *43*, 4797) an Ir complex that effects the coupling of allylic carbonates such as **6** with aromatic amines to give the secondary amine in high ee.

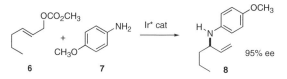

In a complementary approach, Larry Overman of the University of California at Irvine, has developed (*J. Org. Chem.* **2004**, *69*, 8101) a Co catalyst that effects the rearrangement of allylic imidates such as **9** with high ee. There is no need for the starting allylic alcohols to be perfectly trans, as imidates from cis allylic alcohols do not participate in the rearrangement.

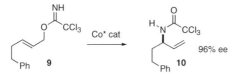

The first-developed method for the enantioselective construction of secondary aminated stereocenters, by Knowles, was the hydrogenation of enamides such as **11**. In the context of a synthesis of BILN 2061 **15**, an antiviral protease inhibitor, Anne-Marie Faucher of Boehringer-Ingelheim, Laval has shown (*Organic Lett.* **2004**, *6*, 2901) that such hydrogenations can be effected even in the presence of a terminal vinyl group. The product **12** was carried on to **15** over several steps, including the Ru (Hoyveda) cyclization of **13** to **14**.

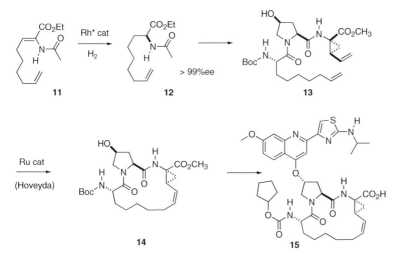

Enantioselective Synthesis of the Polyene Antibiotic Aglycone Rimocidinolide Methyl Ester

August 1, 2005

The complex polyene macrolide antibiotics are clinically effective as antifungal agents. Scott Rychnovsky of the University of California at Irvine has reported (*Angew. Chem. Int. Ed.* **2004**, *43*, 2822) the first synthesis of rimocidinolide methyl ester **4**, the aglycone of rimocidin **1**. The key step in the synthesis is the condensation of the aldehyde **2** with the phosphonate **3**, leading to **4**.

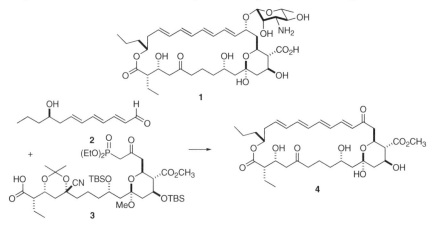

Preparation of the Aldehyde 2: The absolute configuration of the triene aldehyde **2** was set by Noyori hydrogenation of ethyl butyrylacetate **5**. Silylation and Dibal reduction then gave the aldehyde **6**. Reduction of the homologated ester gave the alcohol, which was oxidized to the desired aldehyde **7** by the Swern procedure. Condensation of **7** with the Wollenberg stannyl diene followed by deprotection then gave the unstable aldehyde **2**.

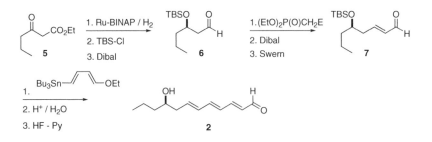

The power of convergent synthesis is illustrated by the preparation of the acid **3**. The three components **9**, **12**, and **17** were each prepared in enantiomerically-pure form using readily-available chiral reagents, followed by functional group manipulation. One of the more remarkable transformations was the homologation of the Weinreb amide **15** to give the unstable allyl ketone, which was then reduced with high diastereoselectivity to give the diol **16**.

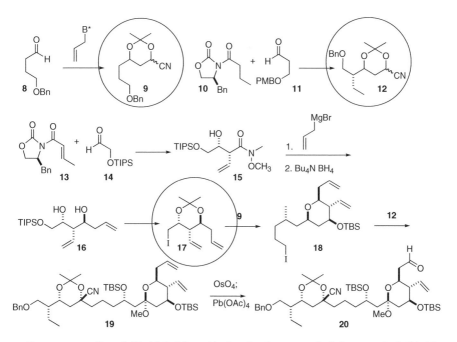

Convergent coupling of **17** with **9**, followed by functional group manipulation, gave the iodide **18**, which was then homologated with **12** to give **19**. Although the two monosubstituted alkenes of **19** appear to be similar, dihydroxylation with OsO_4 was remarkably selective, leading to the aldehyde **20**.

To complete the synthesis of **1**, the acid **3** derived from **20** was converted to the mixed anhydride (Yamaguchi coupling), then esterified with **2**. Exposure to K_2CO_3/18-crown-6 gave the all-*E* tetraene, which was deprotected to give the aglycone **4**.

Enantioselective Transformations of Prochiral Rings

August 8, 2005

Enantiomericallly-pure carbacyclic rings are important components both of physiologically-active natural products and of important pharmaceuticals. Often it is most effective to control the absolute configuration of the ring as it is formed. Recent developments in the enantioselective construction of carbacyclic rings will be covered next week and the week after. The focus this week is on the asymmetric transformation of preformed prochiral rings.

The power of such an approach is illustrated by the synthesis of (-)-pumiliotoxin C **3** recently reported (*Tetrahedron* **2004**, *60*, 9687) by Adriaan J. Minnaard and Ben L. Feringa of the University of Groningen in the Netherlands. Enantioselective conjugate addition of Me$_2$Zn to the inexpensive cyclohexenone **1** gave the intermediate enolate, Pd-mediated coupling of which with allyl acetate gave the ketone **2** in 96% ee. Conversion of the ketone to the cis tosylamide followed by Pd-mediated cyclizative coupling with 1-bromo-1-propene and hydrogenation then gave **3**. The intermediate tosylamide was recrystallized to >99% ee.

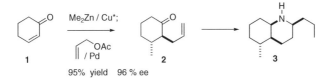

Stephen L. Buchwald at MIT has reported (*Org. Lett.* **2004**, *6*, 4809) a complementary approach. Enantioselective conjugate *reduction* of the inexpensive 3-methylcyclopentenone **4** led to the silyl enol ether **5**, Pd-mediated coupling of which with the aryl halide gave the product **6**. 3-Methylcyclohexenone gave the analogous product in 84% ee.

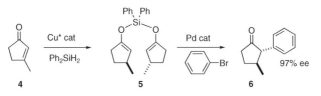

These approaches led to *ternary* stereogenic centers. Brian Stoltz has found (*J. Am. Chem. Soc.* **2004**, *126*, 15044) that the Pd-mediated conversion of **7** to **8** can be induced to proceed in high enantiomeric excess. This appears to be a general method for the preparation of *quaternary* stereogenic centers. Wacker oxidation followed by aldol condensation converted **8** into the bicyclic enone **9**.

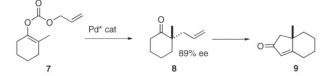

7   8   9

We have put forward (*J. Am. Chem. Soc.* **2004**, *126*, 13900) an alternative approach to the enantioselective construction of cyclic quaternary centers. Addition of phenylacetylene to cyclopentanone followed by dehydration and Shi epoxidation gave the epoxide **10**. Opening of the epoxide with allylmagnesium chloride proceeded with inversion, to give **11**. The alcohol **11** can also be carried on to bicyclic products, exemplified by the sulfone **12**.

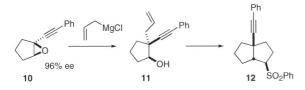

10   11   12

The power of asymmetric transformation of a preformed ring is further illustrated by the synthesis of (+)-allocyathin B₂ **15** reported (*Org. Lett.* **2004**, *6*, 4897) by Masahisa Nakada of Waseda University, Tokyo. This bird nest fungi diterpene was assembled by convergent coupling of enantiomerically-pure **13** with enantiomerically-pure **14**. The cyclohexane **14** was derived from the prochiral cyclohexane-1,3-dione **16**. The cyclic quaternary stereogenic center was established by yeast reduction of **16** followed by diastereoselective reduction of the resulting hydroxy ketone, to give the diol **17**. The two alcohols of **17** were differentiated by selective acetonide formation, giving **18**. The chiral cyclopentane **13** was prepared by enantioselective intramolecular cyclopropanation.

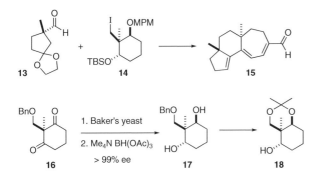

13   14   15

16   17   18

Michael Reactions for Enantioselective Ring Construction

August 15, 2005

An abiding ambition of the organic synthesis chemist is to form a ring in high yield and with high enantiocontrol, using an inexpensive catalyst. The Michael reactions described here come close to achieving that goal.

Intermolecular Michael reactions continue to be developed. Karl Anker Jørgensen of Aarhus University, Denmark, has found (*Angew. Chem. Int. Ed.* **2004**, *43*, 1272) that the organocatalyst **3** mediates the addition of **2** to **1** with high enantiomeric excess. What is more, under the reaction conditions the intial Michael addition is followed by an aldol condensation, to give **4** as essentially a single diastereomer.

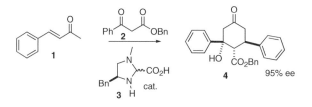

The *intramolecular* Michael reaction is also a powerful transformation. In the cyclizations reported by Tetsuaki Tanaka of Osaka University (*J. Org. Chem.* **2004**, *69*, 6335), the stereochemical outcome is controlled by the chirality of the sulfoxide. Remarkably, subsequent alkylation or aldol condensation leads to one or two additional off-ring stereocenters with high diastereocontrol. Note that the high stereoselectivity in the cyclization is only observed with the (*Z*)-unsaturated ester.

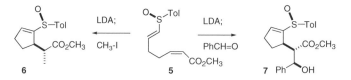

The absolute course of the intramolecular Michael reaction can also be mediated by organo catalysts. Benjamin List of the Max-Planck-Institut, Mülheim, has shown (*Angew. Chem. Int. Ed.* **2004**, *43*, 3958) that **9** is particularly effective. Michael addition followed by intramolecular aldol condensation gives the *trans* bicyclic enone **11** in high ee.

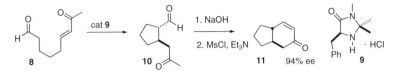

Tandem intramolecular Michael addition – intramolecular alkylation can lead to cyclopropanes. Matthew J. Gaunt of the University of Cambridge has shown (*Angew. Chem. Int. Ed.* **2004**, *43*, 2681) that this intramolecular Michael addition also responds to organocatalysis. In this case, the catalyst, a quinine-derived amine, covalently binds to the substrate, then is released at the end of the reaction.

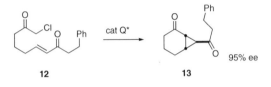

Enantioselective Ring Construction by Intramolecular C-H Insertion and by Cycloaddition

August 22, 2005

The prochiral aziridine **1** is easily prepared from cyclooctene. Paul Müller of the University of Geneva has shown (*Helv. Chim. Acta* **2004**, *87*, 227) that metalation of **1** in the presence of the chiral amine sparteine leads to the bicyclic amine **3** in 75% ee, by way of intramolecular C-H insertion by the intermediate chiral carbene **2**. The sparteine can be recovered and recycled.

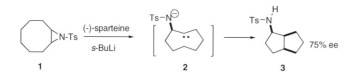

As exemplified by the recent synthesis of (+)-trehazolin **7** by Susumu Ohira of Okayama University of Science (*Tetrahedron Lett.* **2004**, *45*, 7133), carbenes can also be conveniently generated by exposure of ketones such as **4** to lithiated TMS diazomethane. Intramolecular C-H insertion then proceeds with high diastereoselectivity, to give **6**, a key intermediate on the way to **7**.

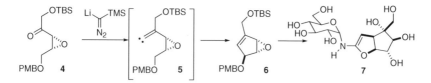

The central role of the Diels-Alder reaction as a workhorse for carbocyclic construction has driven the development of effective chiral catalysts. The state of the art is represented by the catalyst **9**, developed (*J. Am. Chem. Soc.* **2004**, *126*, 13708) by E.J. Corey of Harvard University. This catalyst is effective with a range of dienes and dienophiles, as illustrated by the combination of **8** and **10** to give the endo adduct **11** in high de and ee. The acid corresponding to **11** was a key starting material for the Aubé synthesis of **12**, the Dentrobatid alkaloid **251F**.

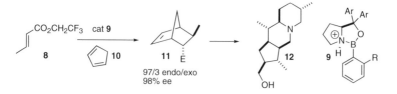

Cyclopentadiene Diels-Alder adducts such as **11** are highly strained. Huw M.L. Davies of the State University of New York, Buffalo, has established (*J. Am. Chem. Soc.* **2004**, *126*, 2692) an important strategic connection between Diels-Alder adducts such as **13** and the enantiomerically-enriched cycloheptane derivative **14**. Oxidative cleavage of the alkene of **14** would lead to a cyclohexane derivative that would be difficult to access by other means. The same adduct **14** (albeit in racemic form) is also available directly by AlCl₃-mediated addition of methacrolein **15** to cyclopentadiene.

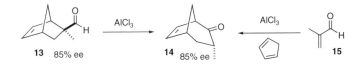

13   85% ee   **14**   85% ee   **15**

Best Synthetic Methods: Construction of Aromatic and Heteroaromatic Rings

August 29, 2005

Aromatic and heteroaromatic rings are the heart of pharmaceutical design. Several useful methods for monocyclic and polycyclic aromatic ring construction have recently been reported.

Aaron L. Odom of Michigan State University has described (*Org. Lett.* **2004**, 6, 2957) a new approach to dialkyl pyrroles. Ti-catalyzed hydroamination of a 1,4-diyne such as **1** leads smoothly to **2**. Similarly, Ti-catalyzed hydroamination of a 1,5-diyne such as **3** delivers **4**. An inherent limitation of this approach is that it only allows substitution at the 2 and the 5 positions of the pyrrole.

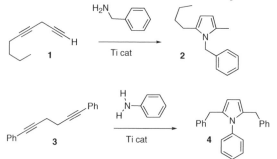

A more flexible approach to pyrroles has been developed (*Tetrahedron Lett.* **2004**, *45*, 9315) by Keiji Maruoka of Kyoto University. Rearrangement of **5** by the Al complex **6** leads, by preferential benzyl migration, to the trisubstituted pyrrole **7**.

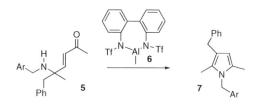

Martin Banwell of the Australian National University has devised (*Org. Lett.* **2004**, 6, 2741) a Pd-mediated approach to fused quinolines. Coupling of **8** with the bromo aldehyde **9** leads to **10**, which on reduction cyclizes to **11**.

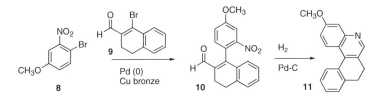

Substituted naphthalenes can be prepared by adding an aromatic ring onto an existing benzene derivative. Rai-Shung Liu of the National Tsing-Hua University, Taiwan, has found (*J. Am. Chem. Soc.* **2004**, *126*, 6895) that exposure of an epoxy alkyne such as **12** to a catalytic amount of TpRuPPh$_3$(CH$_3$CN)$_2$PF$_6$ leads to the β-naphthol **15**. The reaction is thought to be proceeding by initial rearrangement to the vinylidene complex **13**. O-atom transfer converts **13** to the ketene **14**, which undergoes cycloaddition to the newly-liberated alkene to give **15**. The facile conversion of **12** to the TpRu-vinylidene complex **13** will have many other applications in organic synthesis.

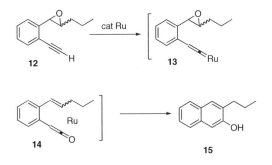

Enantioselective Synthesis of (-)-Epoxomycin

September 5, 2005

For many classes of physiologically-important natural products, total synthesis requires the construction of an extended array of acyclic stereogenic centers, with control of relative and absolute configuration. In the course of a synthesis (*J. Am. Chem. Soc.* **2004**, *126*, 15348) of the proteasome inhibitor (-)-epoxomycin **3**, Lawrence J. Williams of Rutgers University has developed an elegant and potentially powerful solution to this problem. It was apparent that the preparation of **3** reduced to the stereocontrolled synthesis of **2**, since the balance of the natural product is made up of readily-available amino acids. The oxygenated quaternary center of **2** appeared to be a particular challenge. The key insight of the synthesis was that both centers could be established in a single step by selective nucelophilic opening of the enantiomerically-enriched spiro bis epoxide **1**.

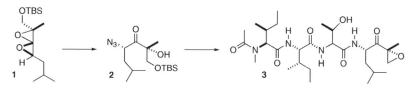

The left hand fragment of epoxymycin **3** was assembled from the previously-described amino acid derivatives **4**, **5**, and **7**, using standard coupling techniques.

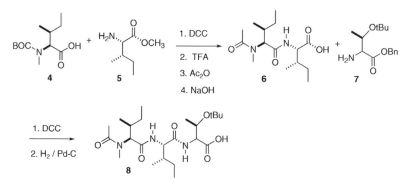

The preparation of the allene bis-epoxide **1** started with isovaleraldehyde **9**. Addition of the protected propargyl alcohol **10** under the Carreira conditions led to **11** in > 95% ee. Mesylation followed by displacement with methyl cuprate provided the allene without loss of enantiomeric excess. Oxidation of the allene **12** with dimethyldioxirane could have led to any of the four diastereomers of the spiro bis epoxide. In the event, only two diastereomers were observed, as a 3:1 mixture. That **1** was the major diastereomer followed from its conversion to **3**. The configuration of the minor diastereromer was not noted. Exposure of **1** to nucleophilic azide then gave the easily-purified **2**.

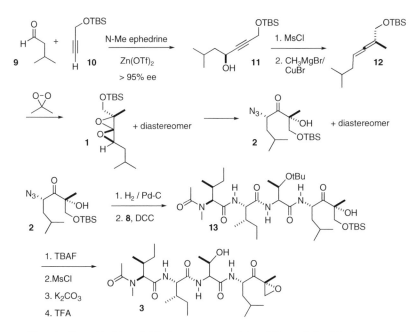

The spirodiepoxide **1** is an intriguing new approach to relative and absolute stereocontrol. It will be interesting to see what other nucleophiles can be used in the opening. It is possible that a chiral oxygen transfer reagent, such as the dioxirane prepared by the Shi protocol, would convert **12** to **1** with improved diastereoselectivity (double diastereoselection).

Best Synthetic Methods: Functionalization of Aromatic and Heteroaromatic Rings

September 12, 2005

Aromatic and heteroaromatic rings are the heart of pharmaceutical design. It is often less expensive to purchase a ring than to build it, so efficient methods for the selective functionalization of preformed aromatic rings are important.

In general, aromatic halogenation proceeds to give the para product. Jallal M. Gnaim of the The Triangle Regional R&D Center, Kfar-Qari, Israel, has developed (*Tetrahedron Lett.* **2004**, *45*, 8471) a procedure that selectively converts phenols such as **1** into the ortho chlorinated product **2**. Meta substitution is even more elusive. William R. Roush, now at Scripps Florida, prepared (*J. Org. Chem.* **2004**, *69*, 4906) the valuable meta brominated phenol **4** by perbromination followed by selective reduction.

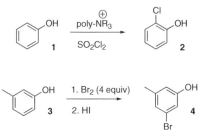

A common application of aryl halides is Pd-mediated coupling to form C-C bonds. Graham Meek of Chirotech Technology Ltd., Cambridge, UK has found (*Tetrahedron Lett.* **2004**, *45*, 9277) that the enamide **6** participates efficiently in Heck coupling. The product **7** is an excellent substrate for enantioselective hydrogenation, to give **8**.

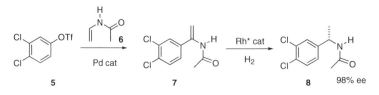

For some applications, it is useful to put a substrate on a solid support. Linkers that can be converted directly to desired functionality ("traceless") are particularly valuable. Andrew M. Cammidge of the University of East Anglia and A. Ganesan of the University of Southampton independently (*Chem. Commun.* **2004**, 1914, 1916) developed the polymeric sulfonyl chloride **9**. The derived phenyl sulfonates are useful partners for transition-metal mediated cross coupling.

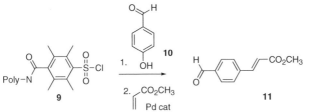

Direct alkylation of an unfunctionalized aromatic ring is usually much less expensive than cross coupling, even with an aryl chloride. This is not, however, usually a useful way to attach a linear chain, as such chains rearrange too readily under Lewis acid conditions. Chuan He of the University of Chicago has found (*J. Am. Chem. Soc.* **2004**, *126*, 13596) an Au catalyst that mediates the direct alkylation of an aromatic ring such as **12** with a straight chain triflate, without rearrangement, to give **13**.

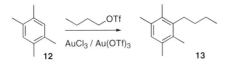

The functionalization of heteroaromatics is also important. Chi-Ming Che of the University of Hong Kong has found (*Organic Lett.* **2004**, *6*, 2405) that in situ oxidation of sulfonamide in the presence of a Ru porphyrin catalyst leads to a species that effects electrophilic amination of heteroaromatics such as **14**, to give **15**.

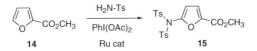

Heterocycles are often sufficiently electron rich that oxidative addition and subsequent coupling can be carried out even on chlorides. Iodides are, nonetheless, much more reactive. Morris Robbins of Brigham Young University has shown (*Organic Lett.* **2004**, *6*, 2917) that Finkelstein exchange of the heteroaryl chloride **16** proceeds readily at –40 °C. The product iodide **17** is an efficient substrate for S$_N$Ar, Sonogashira and Suzuki-Miyaura reactions. For instance, the Sonogashira reaction on **17** proceeded in 20 min at room temperature, to give **18**. Under the same conditions, the chloride **16** did not react.

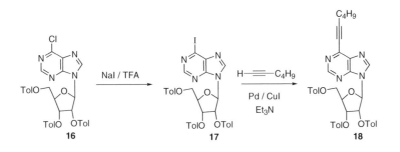

175

Best Synthetic Methods: Oxidation

September 19, 2005

Efficient new methods for oxidation are always welcome. The Dess-Martin periodinane has become the workhorse for alcohol to aldehyde or ketone conversion in organic research labs around the world. Viktor V. Zhdankin of the University of Minnesota, Duluth has described (*Chem. Commun.* **2004**, 106) a complementary family of reagents. Oxidation of an ester **1** of *o*-iodobenzoic acid with NaOCl delivers **2**. Depending on the ester, the reagent **2** is soluble and an effective oxidant, with KBr catalysis, in a wide range of organic solvents. Presumably, the spent oxidant can be recovered and recycled.

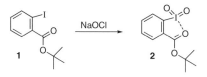

Chromium-based oxidants, noteworthy for their specificity and ease of use, continue to be popular. Enayatollah Mottaghinejad of Azad University of Iran, Tehran has found (*Tetrahedron Lett.* **2004**, *45*, 8823) that barium dichromate, easily prepared, smoothly oxidizes alcohols to aldehydes and ketones in refluxing acetonitrile.

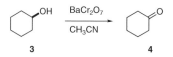

A recent report (*J. Org. Chem.* **2004**, *69*, 8510) by Paul G. Williard of Brown University and Ruggero Curci of Università di Bara of the oxidation of **5** to **6** serves as a timely reminder that the widely-used epoxidation reagent dimethyl dioxirane is also useful for the oxidation of secondary alcohols to ketones.

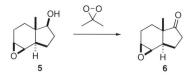

α,β-Unsaturated aldehydes can also be prepared by the net oxidation of allylic halides. Paul B. Jones of Wake Forest University has put forward (*Organic Lett.* **2004**, *6*, 3767) an elegant new method for effecting this transformation, based on nucleophilic displacement by the phenol **8**. Photolysis of the stable aryl ether **9** delivers the unstable aldehyde **10**.

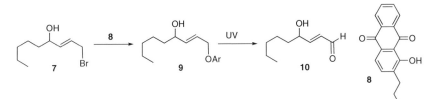

Allylic oxidation offers another route to ketones. Michael P. Doyle of the University of Maryland has found (*J. Am. Chem. Soc.* **2004**, *126*, 13622) that Rh caprolactam is a very active (0.1 mol %) catalyst for this conversion.

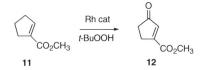

Although the transformation is not an oxidation, Eric N. Jacobsen of Harvard University has developed (*J. Am. Chem. Soc.* **2004**, *126*, 14724) an elegant route to enantiomerically-pure alcohols, based on the conjugate addition of an oxime such as **13** to an α,β-unsaturated imide such as **14**.

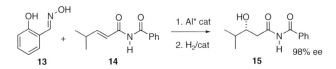

Enantioselective Allylic Carbon-Carbon Bond

September 26, 2005

Although many methods have been developed for enantioselective carbon-carbon construction, not all of these are robust and scalable. A promising recent addition has been the development of protocols for the enantioselective construction of allylic carbon-carbon bonds.

A practical method for the enantioselective addition of an allylic nucleophile to an aldehyde has been acid-mediated allyl transfer, as exemplified by the conversion of **1** and **2** to **3**. While this method worked well for crotyl, allyl transfer itself suffered from eroded ee's. Teck-Peng Loh of the National University of Singapore has found (*Tetrahedron Lett.* **2004**, *45*, 5819) that camphorsulfonic acid (CSA) mediates this conversion without racemization. The alcohol **1** is prepared by addition of allyl Grignard to camphor, so both enantiomers are readily available.

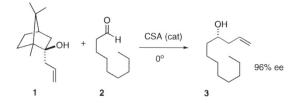

Asymmetric addition of functionalized allylic nucleophiles is also a useful process. Yoshito Kishi of Harvard University has shown (*J. Am. Chem. Soc.* **2004**, *126*, 12248) that 2,3-dibromopropene **4** will add with high enantioselectivity to linear and branched aliphatic aldehydes. The analogous Cl and I derivatives can also be prepared, using the same approach.

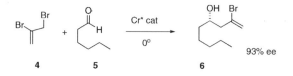

Carbon-carbon ternary centers can also be assembled by nucleophilic addition. Manfred Braun of the Universität Düsseldorf has devised (*Angew. Chem. Int. Ed.* **2004**, *43*, 514) the Lewis acidic Ti complex **8**. Exposure of a racemic benzylic silyl ether such as **7** to allyltrimethylsilane in the presence of the catalyst **8** leads to the alkylated product **9**. Racemic tertiary benzylic ethers are also converted to the alkylated *quaternary* centers in > 90% ee.

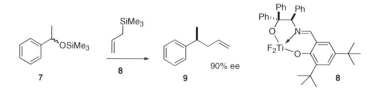

Prochiral allylic leaving groups can also be displaced with high enantioselectivity. Günter Helmchen of the Universität Heidelberg has reported (*Angew. Chem. Int. Ed.* **2004**, *43*, 4595) the optimization of the Ir*-catalyzed malonate displacement of terminal allylic carbonates such as **10**. The reaction proceeds to give predominantly the desired branched product **11**, in high ee.

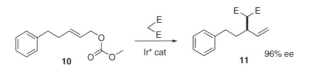

An equivalent transformation can be effected on prochiral allylic chlorides such as **12**, using organometallic nucleophiles. In an important advance, Alexandre Alexakis of the Université de Genève has shown (*Angew. Chem. Int. Ed.* **2004**, *43*, 2426) that using a chiral Cu catalyst, alkyl *Grignard* reagents participate as the nucleophile, giving the product **13** in high regioselectivity and ee.

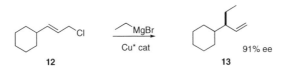

Synthesis of (+)-Cyanthiwigin U

October 3, 2005

The cyanthin diterpenes show physiological activity ranging from cytotoxicity to nerve-growth factor stimulation. Andrew J. Phillips of the University of Colorado recently described (*J. Am. Chem. Soc.* **2005**, *127*, 5334) a concise enantioselective synthesis of cyanthiwigin U **3**, based on the metathesis conversion of **1** to **2**, using the second generation Grubbs catalyst.

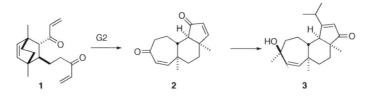

It was clear that **1** would be derived from a Diels-Alder adduct. There has been a great deal of work in recent years around the development of enantioselective catalysts for the Diels-Alder reaction, but the catalysts that have been developed to date only work with *activated* dienophile-diene combinations. For less reactive dienes, it is still necessary to use chiral auxiliary control. One of the more effective of those was the known camphor-derived tertiary alcohol, so that was used in this project. Diels-Alder cycloaddition of the diene **4** with the enantiomerically-pure enone **5** led to the adduct **6** with high diastereocontrol. Oxidative cleavage led to the acid **7**, which was carried on to the bis-enone **1**.

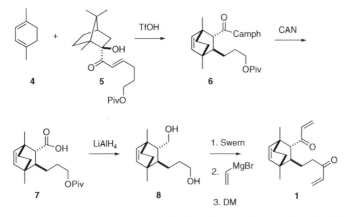

The two-directional tandem metathesis of **1** to **2** proceeded smoothly using 20 mol % of the second generation Grubbs catalyst (now commercially available only from Aldrich and from Materia) under an atmosphere of ethylene. The conversion of **2** to **3** took advantage of the differing reactivity of the two ketones. Addition of hydride to **2** from the less hindered face of the less hindered ketone delivered **4**.

Addition of isopropyl lithium to the surviving ketone followed by oxidative rearrangement of the resulting tertiary allylic alcohol and concomitant oxidation of the secondary allylic alcohol gave the diketone **10**. Selective addition of methyl lithium to the less hindered of the two ketones, again from the more open face, then gave **3**.

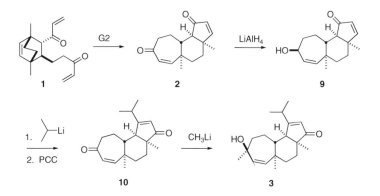

The elegantly concise strategy displayed here for the enantioselective and diastereoselective construction of the tricyclic enone **3**, by two-directional tandem methathesis of the Diels-Alder derived diketone **1**, should have some generality.

Catalysts and Strategies for Alkene Metathesis

October 10, 2005

The Grubbs second generation catalyst (G2) continues to be the workhorse for academic investigations of synthetic applications of alkene metathesis. The requirement by Materia for licensing fees even for research investigations using G2 have made this catalyst much less attractive for industry-based researchers. There is a real interest in the development of alternative catalysts that are robust, active and easily prepared. Pierre Dixneuf of the Université de Rennes has found (*Angew. Chem. Int. Ed.* **2005**, *44*, 2576) that exposure of the arene Ru complex **2** to a propargyl ether such as **3** generates *in situ* a very active metathesis catalyst. The catalyst so generated is apparently an 18-electron species, in contrast to the 16-electron G2. The complex **2**, "a microcrystalline red powder' is prepared by addition of PCy₃ to the commercially-available [(p-cymene)RuCl₂]₂ followed by treatment of the product with AgOTf.

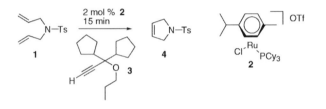

Amir H. Hoveyda of Boston College has reported (*J. Am. Chem. Soc.* **2005**, *127*, 8526) the development of a family of chiral Mo metathesis catalysts that convert prochiral dienes such as **5** and **8**

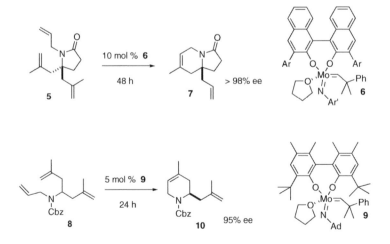

into the cyclized product with high ee. Note that the six examples in the paper that were optimized to ≥ 90% ee required *four different* chiral Mo catalysts. This would not be a concern for manufacturing, where it would be worth the time to find the catalyst that gave the best results.

Several years ago, Professor Hoveyda designed the chelated Ru complex **12a** as a versatile and stable metathesis catalyst. Dennis P. Curran of the University of Pittsburgh has now introduced (*J. Org. Chem.* **2005**, *70*, 1636) the fluorous-tagged Ru catalyst **12b**. The fluorous tag allows the facile recovery of most of the active catalyst. The advantages of this are two-fold: the valuable catalyst can be re-used, and there will potentially be less Ru contamination in the cyclized product.

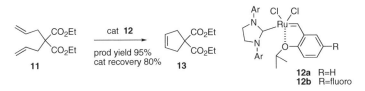

Ring-forming metathesis does not inevitably proceed smoothly. Johann Mulzer of the Universität Wien had planned (*Organic Lett.* **2005**, *7*, 1311) to set the trisubstituted alkene of epithilone by cyclization of the ester **13**. In fact, however, this gave only the undesired dimer. There are two factors that disfavor the cyclization of **13**: the ester prefers the extended rather than the lactone conformation, and the product eight-membered ring would have substantial transannular ring strain. The alternative bis silyl ether **14** does not have such conformational issues - indeed, the buttressing of the dialkyl silyl group *favors* cyclization. Further, the nine-membered cyclic ether product has significantly less transannular strain. Unlike **13**, the bis ether **14** cyclized smoothly.

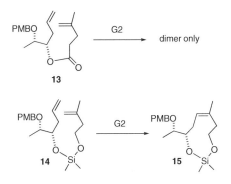

Congratulations to the recipients of the 2005 Nobel Prize in Chemistry! The five Organic Highlights columns for the month of October will be devoted to organic synthesis applications of alkene metathesis.

N-Heterocycle Construction by Alkene Metathesis

October 17, 2005

The first N-heterocycles prepared by alkene metathesis were simple five- and six-membered ring amides. Ring-closing metathesis of free amines is much more difficult. The diene **1**, for instance, gave only low yields of cyclization product. Wen-Jing Xiao of Central China Normal University and Zhengkun Yu of the Dalian Institute of Chemical Physics have shown (*Organic Lett.* **2005** 7, 871) that precomplexation of **1** with the inexpensive Ti(OiPr)$_4$ ties up the amine, allowing for facile cyclization.

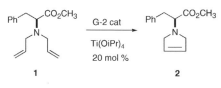

Ring construction by intramolecular alkene metathesis is not limited to five- and six-membered rings. Robert W. Marquis of GlaxoSmithKline, Collegeville, PA faced the challenge (*Tetrahedron Lett.* **2005**, *46*, 2799) of preparing **5**, a potent inhibitor of the osteoclast-specific cysteine protease cathepsin K. The absolute and relative configuration of **3** were established by an Evans aldol condensation. The aldol product **3** could be used directly in the metathesis reaction. Hydrolysis and Curtius rearrangement then led to **5**.

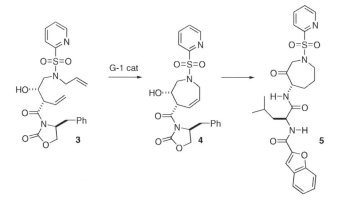

Thermodynamically-difficult medium rings offer a particular challenge for ring-closing metathesis. William D. Lubell of the Université de Montréal has found (*J. Org. Chem.* **2005**, *70*, 3838) that while a *secondary* amide such as **6a** is reluctant to cyclize, the corresponding *tertiary* amide **6b** participates smoothly, leading to **7b**. Tertiary, but not secondary, amides also worked well for nine-membered rings. Secondary and tertiary amides both worked well for ten-membered ring formation.

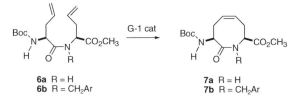

6a R = H
6b R = CH₂Ar

7a R = H
7b R = CH₂Ar

The N-heterocyclic alkenes derived from ring-closing metathesis are useful substrates for further transformation. In a synthesis directed toward the insecticidal cripowellin B **12**, Dieter Enders of RWTH Aachen has shown (*Angew. Chem. Int. Ed.* **2005**, *44*, 3766) that the tertiary amide **8** cyclizes efficiently to the nine-membered alkene **9**. The vision was that an intramolecular Heck cyclization could then deliver the cripowellin skeleton. Indeed, the Heck did proceed, and, depending on conditions, could be directed toward either **10** or **11**. Unfortunately, the conformation of **9** is such that the cyclization proceeded cleanly across the *undesired* face. Nevertheless, both **10** and **11** appear to be valuable intermediates for further transformation.

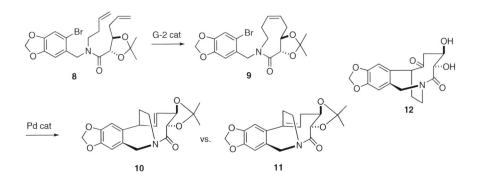

O-Heterocycle Construction by Alkene Metathesis

October 24, 2005

In the preparation of enantiomerically-pure starting materials, it is not uncommon for the early low molecular weight intermediates to require special handling. Often, the initial stereogenic centers are derived from carbohydrate precursors. Bastien Nay of the Muséum National, d'Histoire, Paris has developed (*Tetrahedron Lett.* **2005**, *46*, 3867) an elegant approach that takes advantage of alkene cross metathesis. Enantiomerically-pure diols such as **1**, readily prepared from mannitol, are easy to handle. Exposure of the derived acrylate to the second generation Grubbs catalyst gives clean transformation to two equivalents of the γ-lactone **3**. The corresponding δ-lactones are formed even more efficiently from the vinyl acetate esters of **1**.

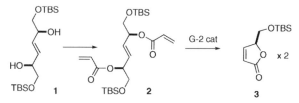

Kevin J. Quinn of the College of the Holy Cross chose (*Organic Lett.* **2005**, *7*, 1243) a complementary approach in his synthesis of rollicosin. The symmetrical diol **4** is also available from carbohydrate precursors. Monosilylation followed by esterification with acryloyl chloride gave **5**. Exposure of **5** to the Grubbs catalyst in the presence of **6** led, by ring-closing metathesis and cross metathesis, to the γ-lactone **7**. Note that δ-lactone formation did not compete!

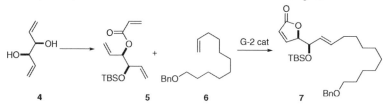

En route to a synthesis of (+)-peluroside A, Mikhail S. Ermolenko of the Institut de Chemie, Gif-sur-Yvette, envisioned (*Organic Lett.* **2005**, *7*, 2225) selective construction of the trisubstituted alkene by ring-closing metathesis of the ester **8**. Unfortunately, preparation of **8** was accompanied by substantial racemization, to give **9**. In fact, this proved to be an advantage, because it led to the observation that **8** participated in ring-closing metathesis about 50 times as rapidly as **9**, delivering **10** contaminated with only a trace of the diastereomeric δ-lactone. Thus, the mixture of **8** and **9** was prepared using the racemic acid choride, and the unreacted **9** was recycled to the mixture of **8** and **9** by brief exposure to LDA.

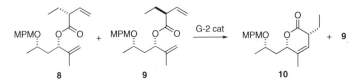

Stereodefined spiroketals are a common structural motif in physiologically-active natural products. Richard P. Hsung of the University of Minnesota recently reported (*Organic Lett.* **2005**, 7, 2273) that Tf₂NH is a particulary effective Brønsted acid mediator for the stereoselective coupling of vinyl lactols such as **11** with homoallylic acids such as **12**. The axial ethers so produced undergo smooth ring-closing metathesis to the spiroketals.

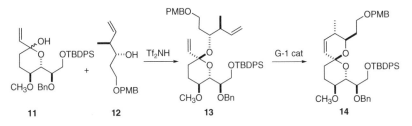

Ring-closing metathesis can also be used to close larger rings. In the course of a total synthesis of (-)-dactylolide, Michael P. Jennings of the Unversity of Alabama described (*Organic Lett.* **2005**, 7, 2321) the preparation of the diol **15**. Despite the presence of four other alkenes in **15**, brief exposure to the second generation Grubbs' catalyst (1 h, rt) delivered **16** in excellent yield. Both of the diastereomers of **15** appeared to participate with equal efficiency in the cyclization.

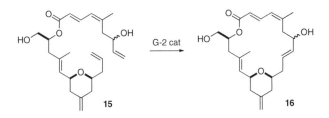

Alkene Metathesis in Total Synthesis: Valienamine, Agelastatin and Tonantzitlolone

October 31, 2005

Applications to natural product synthesis probe the limits of any synthetic method, as situations arise that would never have been considered in the course of first developing the method. Recent syntheses of valienamine **6**, agelastatin **9** and tonantzitlolone **15** pressed the limits of ring-closing metathesis.

The efficient conversion of carbohydrates to highly-oxygenated, enantiomerically-pure carbocycles has long been a goal of organic synthesis. Heung Bae Jeon and Kwan Soo Kim of Yonsei University, Seoul, have optimized the conversion (*J. Org. Chem.* **2005**, *70*, 3299) of the commercial 2,3,4,6-tetra-O-benzyl-D-glucopyranose **1** to the alkene **3**. Hydrolysis of the thioacetal **3** delivered an unstable aldehyde, that was reacted with vinyl magnesium bromide to give **4** as an inseparable 7:3 mixture of diastereomers. The highly-oxygenated diene **4** participated smoothly in the Grubbs cyclization, to give the easily-purified alcohol **5** as the major product. The alcohol **5** was carried on to the *a*-glucosidase inhibitor valienamine **6**.

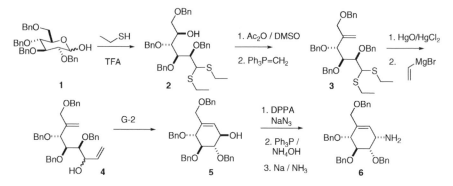

Highly aminated substrates also participate smoothly in the Grubbs cyclization. Franklin A. Davis of Temple University has reported (*Organic Lett.* **2005**, *7*, 621) the preparation of the enantiomerically-pure 1,2-diamine **7** using his elegant sulfinimine addition. Remarkably, the highly-aminated diene **7** participated smoothly in the Grubbs cyclization. The product **8** was carried on to the novel antitumor agent (-)-agelastatin A **9**, the central cyclopentane ring of which is unusual in having four aminated stereogenic centers.

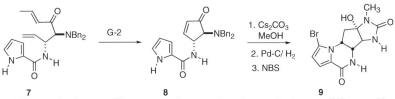

7   8   9

The vagaries that can afflict any complex natural product synthesis are well illustrated by the route to the cytotoxic diterpene tonantzitlolone **15** developed (*Organic Lett.* **2005**, *7*, 479) by Andreas Kirschning of the Universität Hannover. Their first approach, based on ring-closing metathesis to form the single alkene of **15**, had to be abandoned, because metathesis failed, probably because the product alkene was too congested. This led them to the approach illustrated, the expectation being that if the other alkene was too congested to be *formed* by ring-closing metathesis, it would be too congested to *interfere* with ring-closing metathesis. Aldol addition of **10** proceeded with remarkably high diastereoselectivity, leading to **11**. Ring-closing metathesis of the keto diol went well, setting the stage for the preparation of **15**. The synthetic material so prepared turned out to be the enantiomer of the natural product.

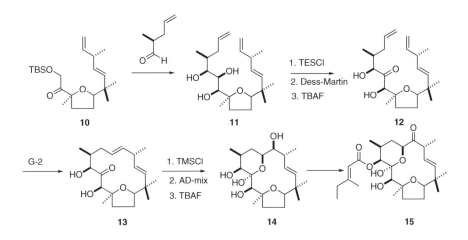

10   11   12

13   14   15

Total Synthesis of the Tetracyclines

November 7, 2005

Although the tetracycline antibiotics have been mainstays of antibacterial chemotherapy for decades, they had eluded efficient total synthesis. In a landmark accomplishment, Andrew G. Myers of Harvard University recently reported (*Science* **2005**, *308*, 395; *J. Am. Chem. Soc.* **2005**, *127*, 8292) the first such syntheses.

In the *Science* paper, which was published first, total syntheses of the clinically-important 6-deoxytetracyclines, including doxycycline **9**, are described. The starting point for the synthesis was the enantiomerically-pure ester **2**, prepared by fermentation of benzoic acid **1** to the 1,2-dihydrodiol, followed by epoxidation, rearrangement and silylation. Acylation of **3** with **2** gave the ketone **4**, which on exposure to LiOTf underwent a very interesting, and diastereoselective, carbon-carbon bond forming reaction to give, after selective desilylation with TFA, the alcohol **5**. The authors speculate that this reaction is proceeding by initial S_N2' epoxide opening by the N, followed by ylide formation and 2,3-rearrangement.

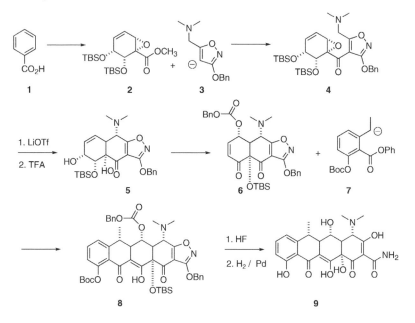

The alcohol **5** was the common intermediate for both syntheses. For the deoxy series, **5** was carried on to the enone **6**. Conjugate addition of the anion **7** proceeded with remarkable diastereoselectivity, to give, after intramolecular acylation and deprotection, doxycycline **9**.

The *JACS* paper describes the total synthesis of the more highly oxygenated (-)-tetracycline **16**. To this end, the alcohol **5** was carried on to the enone **10**. Opening of the cyclobutane **11** to the *o*-quinone methide followed by Diels-Alder cycloaddition to **10** delivered the endo adduct **12**.

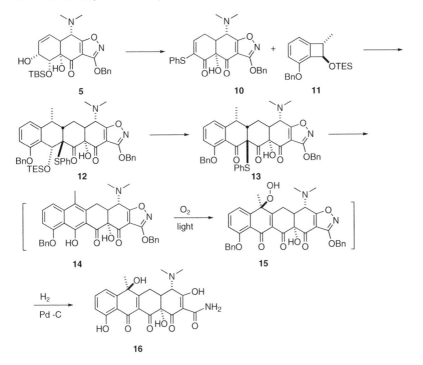

Deprotection and oxidation of **12** gave **13**, which was further oxidized to the sulfoxide. Elimination of the sulfoxide gave the naphthalene derivative **14**, which underwent spontaneous oxidation to **15**. Reductive deprotection then gave tetracycline **16**. The diastereoselectivity of the air and light-mediated oxidation is remarkable.

Enantioselective Construction of N-Heterocycles

November 14, 2005

With increasing emphasis on the single-enantiomer synthesis of pharmaceuticals, there is a need for efficient methods for the preparation of enantiomerically-enriched N and O heterocycles.

Where possible, it may be most economical to effect a chiral transformation on a pre-formed, pro-chiral ring. Ben Feringa of the University of Groningen prepared (*Chem Commun.* **2005**, 1711) the enone **2** from 4-methoxypyridine **1**. Cu*-catalyzed conjugate addition of dialkyl zincs to **2** proceeded in up 96% ee. Pd-mediated allylation of the intermediate zinc enolate led to **3**, with the two alkyl subsituents exclusively trans to each other.

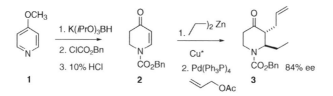

Enantiomerically-enriched piperidines can also be prepared by hydrogenation of pyridine derivatives. André Charette of the Université de Montréal found (*J. Am. Chem. Soc.* **2005**, *127*, 8966) that ylids such as **5**, prepared directly from the pyridine **4**, gave the highest ee's on Ir*-catalyzed hydrogenation. An advantage of this approach is that the piperidine derivatives **6** are crystalline, and are easily recrystallized to higher ee.

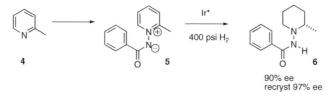

Mercedes Amat and Joan Bosch of the University of Barcelona have been exploring (*Chem. Commun.* **2005**, 1327) a kinetic resolution route to piperidines. Condensation of a ketone or aldehyde ester such as **7** with an enantiomerically-pure amino alcohol such as **8** with proceeds with high (15:1) diastereoselectivity, to give **9**. Reduction of **9** then delivers the piperidine **10** in high enantiomeric excess.

192

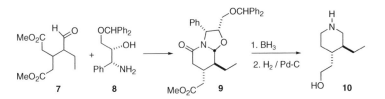

Polyhydroxylated piperidines such as **16** are of interest as glucosidase inhibitors. Antoni Riera, also of the University of Barcelona, has developed (*J. Org. Chem.* **2005**, *70*, 2325) a route to **16** from the readily-available enantiomerically-pure epoxide **11**. Condensation with allyl isocyanate **12** followed by cyclization gave **13**, which was further cyclized by a Grubb's catalyst (unspecified) to **14**. Protection set the stage for face-selective dihydroxylation, to give **15**. Several other piperidines having other polyhydroxylation patterns were also prepared from **14**.

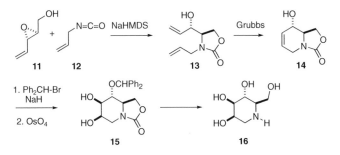

Joseph P.A. Harrity of the University of Sheffield has reported (*J. Org. Chem.* **2005**, *70*, 207) a complementary approach to enantiomerically-pure piperidines. Alkylated azridines such as **17** are readily available from aspartic acid. Pd-catalyzed condensation of **17** with the Trost reagent **18** was found to be most effectively mediated by bis-phosphines such as "dppp", 1,3-bis-diphenylphosphinopropane. The piperidine **19** was the key intermediate for the preparation of several of the *Nuphar* alkaloids, including **20**.

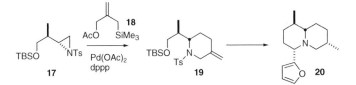

Stereocontrolled Construction of Cyclic Ethers

November 21, 2005

While N-heterocycles have dominated pharmaceutical synthesis from its inception, the increasing facility with which complex cyclic ethers can be assembled suggests that such cyclic ethers could also be suitable pharmaceutical platforms.

Martin E. Fox of Dowpharma Chiratech in Cambridge, UK, has described (*J. Org. Chem.* **2005**, *70*, 1227) a concise route to enantiomerically-pure tetrahydrofuran derivatives such as **3**, starting from methyl D-malate. The alkylation of **1** proceeded with 20:1 diastereoselectivity, leading to the lactone **2**. Reduction followed by exposure to Me$_3$SiBr activated the lactol for coupling with the alkenyl cuprate, to give **3**, again with high diastereoselectivity. The ether **3** is of interest as a prostaglandin analogue.

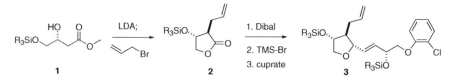

Sugars are also useful starting materials for the preparation of cyclic ethers. Maarten H.D. Postema, now at the Josephine Ford Cancer Center, Detroit, prepared (*J. Org. Chem.* **2005**, *70*, 829) the ester **4** from D-galactal. Methylenation of the ester carbonyl followed by ring-closing metathesis returned **5**. Hydroboration of the enol ether then delivered the C-glycoside **6** in high diastereomeric purity.

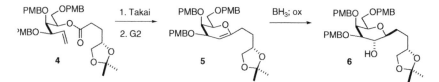

Cyclic enol ethers such as **8** are also easily epoxidized. R. Daniel Little of the University of California, Santa Barbara has found (*J. Org. Chem.* **2005**, *70*, 5249) that such an epoxide is reduced with Ti(III) regioselectively to the radical, that adds with remarkable diastereocontrol to enones such as **7** to give the adduct **9**. Reductive cyclization converted **9** to the tricyclic ether **10**. The C-Br bond of **10** was stable both to the Et$_3$SiH conditions, and to the free radical removal of the xanthate derived from the alcohol.

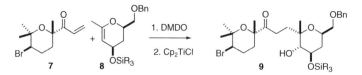

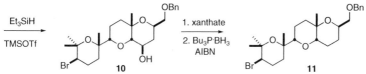

Medium rings can also be formed by ring-closing metathesis. Kenshu Fujiwara of Hokkaido University has demonstrated (*Tetrahedron Lett.* **2005**, *46*, 3465) that the ester **12** undergoes Claisen rearrangement to give **13** as a single diastereomer. Exposure to the second-generation Grubbs catalyst then delivers the eight-membered ring ether **14**.

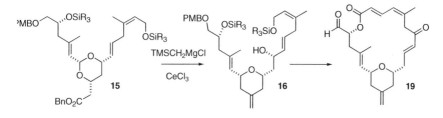

(-)-Zampanolide shows IC$_{50}$ values of 1-5 nanomolar against several cell lines. The immediate precursor to zampanolide is dactylolide **19**. Paul Floreancig of the University of Pittsburgh has developed (*Angew. Chem. Int. Ed.* **2005**, *44*, 3485) a powerful approach to the macrocyclic ring system of **19**. Attempted homologation of **15** to the allyl silane led directly to the cyclic ether **16**, with high diastererocontrol. The stereodefined six-membered ring of **16** then served as a scaffold for assembling the macrolactone **19**.

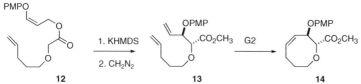

Synthesis of the Proteasome Inhibitors Salinosporamide A, Omuralide and Lactacystin

November 28, 2005

The structurally-related γ-lactams salinosporamide A **1**, omuralide **2** and lactacystin **3**, of bacterial origin, inhibit proteasome activity, and so are of interest as lead compounds for the development of anticancer agents. Barbara C. M. Potts of Nereus Pharmaceuticals in San Diego has reported (*J. Med. Chem.* **2005**, *48*, 3684) a detailed structure-activity studies in this series, and E.J. Corey of Harvard University has prepared (*J. Am. Chem. Soc.* **2005**, *127*, 8974, 15386) several interesting structural analogues. Susumi Hatakeyama of Nagasaki University, building on previous work in this area, has reported (*J. Org. Chem.* **2004**, *69*, 7765) a synthesis of **2** and **3** from Tris.

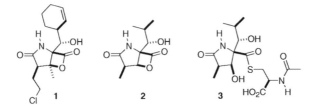

In cell culture, **1** is by far the most active of these three natural products. The challenge in the synthesis of **1** is not closing the β-lactone, but rather the stereocontrolled assembly of the γ-lactam **9**. E.J. Corey reported (*J. Am. Chem. Soc.* **2004**, *126*, 6230) the first route to **1**. The acrylamide **5** was prepared from (*S*)-threonine methyl ester. Highly diastereoselective (9:1) intramolecular Baylis-Hillman

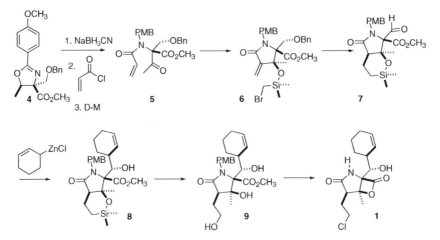

196

condensation of **5** followed by silyation led to **6**, which cyclized under Stork conditions to the cis-fused **7**. Addition of cyclohexenyl zinc proceeded with remarkable diastereocontrol, to give **8**.

Samuel Danishefsky of Columbia University has also described (*J. Am. Chem. Soc.* **2005**, *127*, 8298) a total synthesis of **1**, starting from the pyroglutamate derivative **10**. Conjugate addition followed by alkylation established the lactam framework. Intramolecular cyclization of **12** gave **13**, establishing the aminated quaternary center. The oxygenated quaternary center was then constructed by phenylselenyl-mediated cyclization of **14**. The end game of this synthesis used the already-established cyclohexenyl zinc addition, which worked as well with **16** as it had with **7**.

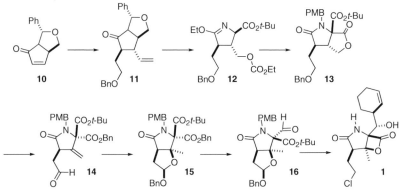

E. J. Corey recently described (*Organic Lett.* **2005**, *7*, 2699, 2703) an even more elegant approach to the γ-lactam. Exposure of **5** to the Kulinkovich conditions followed by iodination delivered **18**, presumably by way of the titanacycle **17**. Although all-carbon trans-5,5 systems are strained, the trans ring fusion is the expected stereochemical outcome with (RO)$_2$Ti in the ring. Complexation of the ester O to the Ti may explain the observed facial selectivity. Brief exposure of **18** to Et$_3$N followed by silylation converted it to **6**. The preparation of several more analogues in this series is also reported in these two articles.

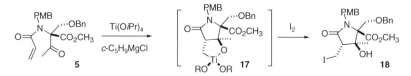

Synthesis of (-)-Sordaricin

December 5, 2005

Sordaricin **2** is the aglycone of sordarin **3**, the parent of a family of clinically-effective antifungal agents. Lewis N. Mander of the Australian National University has published (*J. Org. Chem.* **2005**, *70*, 1654) a full account of their enantioconvergent (see below) synthesis of **2**, based on the intramolecular Diels-Alder cyclization of the triene **1**. For a complementary total synthesis by Narasaka of sordaricin **2**, see *Org. Chem. Highlights* **2005**, April 4.

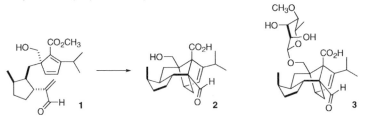

To prepare **1**, two enantiomerically-pure pieces were needed, **10** and **11**. As it developed, **10** was prepared from one enantiomer of **4**, and **11** was prepared from the other enantiomer of **4**. In both the synthesis of **10** and the synthesis of **11**, the folded nature of the tricyclic ring system was used to control relative (and absolute) configuration around the cyclopentane ring. Conjugate addition to (+)-**4** followed by alkylation led to the cis dialkyl **5**. The retro Diels-Alder unmasking of the enone was effected by heating to reflux in 1,2-dichlorobenzene while a stream of nitrogen was bubbled through the solution. Conjugate addition to **6** proceeded to the more open face of the enone, to give **7**. Removal of the carbonyl under conditions that did not epimerize the adjacent methyl group followed by oxidation of the isopropenyl group then led to **10**.

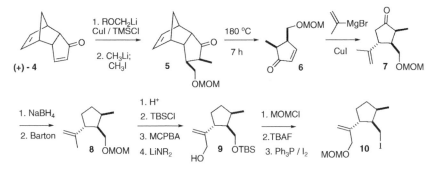

With the iodide **10** in hand, the preparation of the triene **15** could proceed. The single quaternary center of **15** was constructed by deprotonation of the nitrile **11**, followed by alkylation with **10**. Again, bond formation proceeded on the outside face of the tricyclic ketone, to give **12**. Because it was thought (correctly, as it developed) that the Diels-Alder cyclization would proceed with higher selectivity at lower temperature, **13** was cracked to give the enone **14**. Methoxycarbonylation with the Mander reagent followed by enol triflate formation and Cu-mediated coupling then delivered the triene **15**.

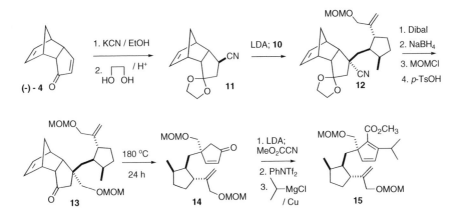

Deprotection of **15** followed by selective oxidation gave the aldehyde **1**. In the event, cyclization of **1** proceeded smoothly, albeit a little slowly, near room temperature, to give exclusively the desired regioisomer. Demethylation then gave (-)-sordaricin **2**.

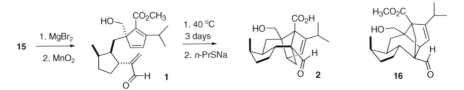

It is interesting that on heating **1** or the cycloadduct above 150 °C, the predominant product (3~4:1) from the cycloaddition is the undesired regioisomer **16**. So, even though the transition state in the intramolecular Diels-Alder cyclization is product-like, the *more* stable transition state for the cyclization of **1** in fact leads to the thermodynamically *less* stable regioisomer **2**.

Recent Advances in Carbocyclic Ketone Construction

December 12, 2005

As the ketone is the central functional group of organic synthesis, so cyclic ketone construction has dominated carbocyclic construction. Several powerful methods for cyclic ketone construction have recently been put forward.

Several procedures are known, including Rh catalysis, for the exo cyclization of alkenyl aldehydes such as **1** to ketones such as **2**. Kiyoshi Tomioka of Kyoto University recently reported (*J. Org. Chem.* **2005**, *70*, 681) that efficient cyclization can be achieved by merely heating the aldehyde with a catalytic amount of a tertiary thiol, with AIBN as the initiator. Addition takes place even to unactivated alkenes, and both five- and six-membered ring ketones can be formed. It is a measure of the mildness of the method that cyclization of **1** gives **2** as a 1:1 mixture, even though trans is much the more stable diasteromer.

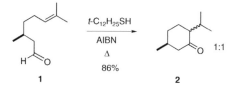

Cyclic ketones can also be formed by intramolecular aldol condensation. Roger C. Whitehead of the University of Manchester found (*Tetrahedron Lett.* **2005**, *46*, 2803) that the cis ene dione **4**, available by oxidation of the corresponding furan **3**, underwent highly diasteroselective aldol condensation, to give untenone A **5**.

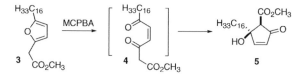

The intramolecular aldol condensation can also be used to prepare enantiomerically-pure cyclic enones. Shigefumi Kuwahara of Tohoku University has shown (*Tetrahedron Lett.* **2005**, *46*, 547) that

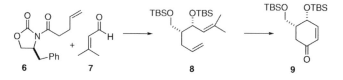

addition of **6** to **7**, following the Evans procedure, followed by reduction and silylation delivered the bis-ether **8**. Wacker oxidation followed by ozonolysis and exposure to aqueous NaOH in ether then led to the enone **9**, without epimerization of the γ-center.

Scott G. Nelson of the University of Pittsburgh has developed (*J. Org. Chem.* **2005**, *70*, 4375) a highly diastereocontrolled route to substituted cyclohexanones using the intramolecular Sakurai reaction. The requisite allyl silane **12** was prepared by Claisen rearrangement of the allylic alcohol **10**, followed by homologation. The Ti enolate from the Sakurai addition was trapped with isobutyraldehyde to give **13**. Although 32 diastereomers of **13** are possible, the diastereomer illustrated was the dominant product from the cylization. Note that use of the enantiomerically-pure form of the alcohol **10** would have led to enantiomerically-pure **13**.

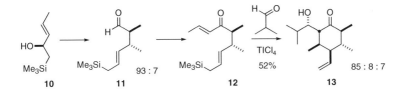

One of the most powerful methods for bicyclic ketone construction is the intramolecular Pauson-Khand reaction (**14** → **15**). Although catalytic methods for this transformation have been put forward, they are not always successful. Jihua Chen and Zhen Yang of Peking University have now found (*Organic Lett.* **2005**, *7*, 593) that the cyclization proceeds quickly and efficiently with 5 mol % of the commercial grade of $Co_2(CO)_8$ if it is run in the presence of the inexpensive tetramethylthiourea. The authors have also reported (*Organic Lett.* **2005**, *7*, 1657) that TMTU is beneficial to the Pd-catalyzed version of the reaction. These advances will make the Pauson-Khand cyclization a more generally practical procedure.

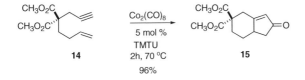

Stereoselective Construction of Carbocyclic Rings

December 19, 2005

Carbocyclic rings, unlike saturated heterocyclic rings, are not susceptible to the hepatic activation that is the basis of much drug toxicity. To pave the way for pharmaceutical discovery based on carbocyclic agents, much effort is going into the development of practical procedures for their stereocontrolled construction.

Intramolecular alkylation is an attractive method for ring construction. Helena M.C. Ferraz of the University of São Paulo and Marcos N. Eberlin of the State University of Campinas report (*J. Org. Chem.* **2005**, *70*, 110) that the intramolecular displacement of the iodide **1** proceeds by way of the tethered enamine **2**, delivering the quaternary center of the keto ester **3** with high diastereocontrol.

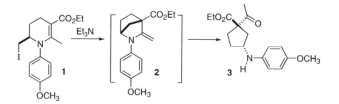

Günter Helmchen of the Ruprecht-Karls-Universität, Heidelberg has reported (*Chem. Commun.* **2005**, 2957) the design of chiral ligands that direct the absolute sense of the Ir-catalyzed intramolecular alkylation of **4** to **5**. Both 5- and 6-membered rings can be prepared in high ee using this procedure. The use of Ir, which is much less expensive than Pd or Rh, is particularly attractive.

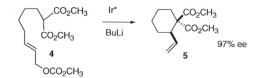

The intramolecular aldol reaction can also proceed with high enantioselectivity. Daniel Romo of Texas A&M University has shown (*J. Org. Chem.* **2005**, *70*, 2835) that a quinine-derived organocatalyst Q^+ works well for the cyclization of aldehyde acids such as **6** to the bicyclic β-lactone **8**. Both 5- and 6-

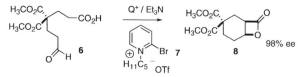

membered rings are formed with high enantioselectivity. The key to the success of the cyclization was the use of the modified Mukaiyama activating agent **7**, having a non-nucleophilic counterion.

Tomislav Rovis of Colorado State University has reported (*Angew. Chem. Int. Ed.* **2005**, *44*, 3264) the diastereoselective Lewis acid-mediated 1,3-rearrangement of dihydrooxepins such as **10** to the cyclopentene carboxaldehyde **11**. It is particularly convenient that the precursor aldehyde **9** is also converted into **11** under the same conditions. As there are many ways to construct alkenyl cyclopropanes such as **9** with control of both relative and absolute configuration, this is an important addition to methods for the stereocontrolled construction of cyclopentanes,

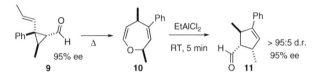

The internal Diels-Alder cycloaddition has long been a workhorse for the construction of bicyclic systems, with substituents on the bridge between the diene and the dienophile directing the stereochemical course of the cyclization. In conjunction with a total synthesis of (+)-dihydrocompactin **16**, Terek Sammakia of the University of Colorado has reported (*J. Am. Chem. Soc.* **2005**, *127*, 6504) the elegant use of the *distal* oxygenated stereogenic center of **12** to direct the stereochemical course of the ring formation. Since lactone formation by the SmI$_2$ Molander protocol is highly diastereocontrolled, the initial oxygenated center of **12** is used to set *every* other stereogenic center in **16**.

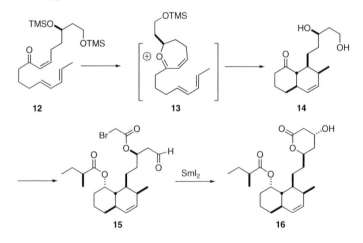

Asymmetric Transformation of Prochiral Carbocyclic Rings

December 26, 2005

The push toward enantiomerically-pure carbocyclic intermediates has led to the development of new methods for the enantiodifferentiation of inexpensive prochiral cyclic starting materials. For instance, Robert H. Morris of the University of Toronto recently reported (*Organic Lett.* **2005**, *7*, 1757) that a family of enantiomerically-pure Ru complexes originally developed for asymmetric transfer hydrogenation also mediate the enantioselective addition of malonate to cyclohexenone.

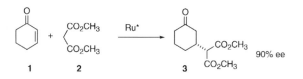

Asymmetric conjugate addition has also been effected with organometallic reagents. Much of the work to date has been with dialkyl zincs, with some reports of Grignard reagents. Alexandre Alexakis of the University of Geneva and Simon Woodward of the University of Nottingham have now described (*Chem Commun.* **2005**, 2843) the use of trialkyl aluminum reagents, with chiral Cu catalysis. Some organoaluminum reagents, such as Me_3Al, are commercially available and easy to handle. Organoaluminum reagents can also be prepared from the alkyne or the alkene.

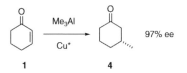

Enantioselective conjugate addition is not limited to 5- and 6-membered rings. Ben L. Feringa and Adriaan J. Minnaard of the University of Groningen describe (*Chem Commun.* **2005**, 1387) the use of the cross-conjugated dienone **5**. The initial addition proceeds with high ee, to give **6**. By choice of the proper absolute configuration of the catalyst, the enone **6** was then carried on selectively to either **7** or **8**.

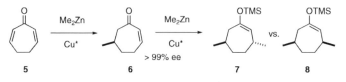

Enantioselective conjugate addition to construct a cyclic quaternary center has been a particular challenge. Alexandre Alexakis has also shown (*Angew. Chem. Int. Ed.* **2005**, *44*, 1376), as illustrated by the conversion of **9** to **13**, that Cu*-catalyzed organoaluminum reagents work effectively in this context.

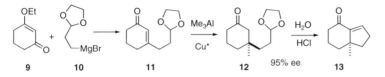

Defined quaternary centers can also be constructed α to ketones, by enantioselective alkylation. Eric N. Jacobsen of Harvard University has found (*J. Am. Chem. Soc.* **2005**, *127*, 62) that tin enolates work particularly well with his Cr salen catalyst. A variety of activated alkylation agents give high ee from the alkylation. It works well for 5-, 6- and 7-membered rings.

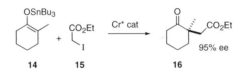

Karl Anker Jorgensen of Aarhus University, Denmark has found (*J. Am. Chem. Soc.* **2005**, *127*, 3670) that face-selective addition can also be carried out on an activated aromatic ring. In this work, the chirality is delivered by a quinine-derived organocatalyst.

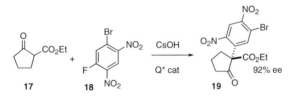

Author Index

Reaction Index